REPRODUCTION IN EUTHERIAN MAMMALS

MARK S. KOSCO

KOSCO PRESS
ISBN 0-9718156-1-5

Correspondence or comments concerning this manuscript should be addressed to

Kosco Press
P.O. Box 954
Saegertown, PA 16433

ISBN 0-9718156-1-5

TABLE OF CONTENTS

PAGE

TABLE OF CONTENTS- CONTINUED

PAGE

TABLE OF CONTENTS – CONTINUED

TABLE OF CONTENTS – CONTINUED

PAGES

TABLE OF FIGURES

PAGES

TABLE OF FIGURES-CONTINUED

PAGE

PAGE

TABLE OF TABLES

SEX AND REPRODUCTION

INTRODUCTION

To the common layman, **sex** is often used in reference to mammalian genital function and reproduction (replication). However, sex is the exchange of genetic material between organisms in which reproduction may or may not occur.

There are three types of sex that occur among living organisms. Transgenic sex occurs among bacteria and was the first form of sex on this planet. Cellular symbiogenesis evolved after transgenic sex and was partially responsible for the development of the eukaryotic cell. Meiotic sex occurs when diploid organisms produce haploid propagules. The haploid propagules then recombine to form a diploid state. Each sex type is discussed below.

TRANSGENIC SEX

Types of Transgenic Sex

Transgenic sex was the first sex to develop on this planet. Transgenic sex is the exchange of genetic material between two organisms, usually bacteria. The types of transgenic sex observed among bacteria include transformation, conjugation, lysogeny, transduction, and phage transfer.

Transformation is the transfer of genetic material from a dead donor bacterium to a live recipient bacterium (Figure 1.1). The study of transformation can be dated to the 1920's when scientist first observed the process while working with the bacterium *Streptococcus pneumoniae* (this bacterium causes pneumonia). When this bacterium is found in nature, it is encapsulated in polysaccharide and causes death upon injection. Mutants devoid of the capsule can also be isolated and are nonlethal. In scientific studies of the 1920's, mice were divided into three groups. The first group was inoculated with nonencapsulated bacteria. The second group was injected with heat killed encapsulated bacteria. The third group was injected with nonencapsulated bacteria and heat killed bacteria. The mice in the first two groups did not develop pneumonia. However, the mice in group three developed pneumonia and died. Scientist were able to isolate encapsulated bacteria from the dead mice of group three. Later research showed that encapsulated bacteria added to nonencapsulated bacteria *in vitro* produced encapsulated cells. In hindsight, mutation, transformation and selection can explain these observations. The heat killed encapsulated bacteria carried the genetic information for the synthesis of the capsule. The mutant bacteria had a defective gene that did not allow the capsule to be produced. Even though the encapsulated bacteria were heat killed and then macerated, their DNA remained intact. The DNA of the encapsulated bacteria then passed through the plasma membrane and cell wall of certain nonencapsulated recipient bacteria. These nonencapsulated bacterial cells then integrated the genes for capsule formation into their own DNA, thus becoming virulent. These newly transformed bacteria were resistant to the immune mechanisms of the mouse and therefore multiplied. Nonencapsulated bacteria were selectively killed by the immune systems of their hosts, thus causing a shift in bacterial population in favor of the encapsulated bacteria. The mice died of pneumonia when the encapsulated bacterial cells reach a critical concentration.

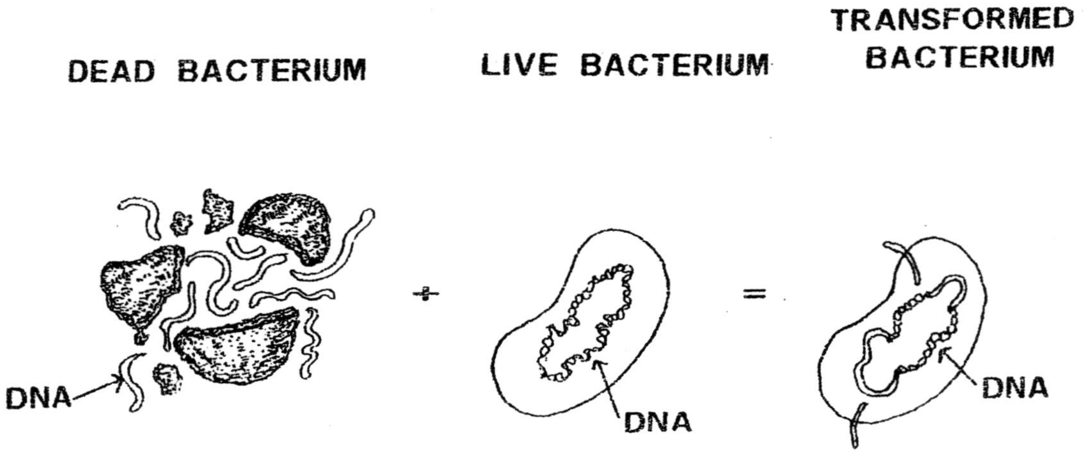

Figure 1.1.   Transformation. DNA from a dead bacterium is transferred and incorporated into the DNA of a recipient live bacterium.

Conjugation is the exchange of genetic material after contact between two bacteria (Figure 1.2). For example, there are two types of cells in a population of *E. coli* for which conjugation occurs. One population, called F+, has a few pieces of extrachromosomal DNA (referred to as F particles) in their cytoplasm that code for specific proteins. Cells missing extrachromosomal DNA are referred to as F-. When an F+ and F- cells are brought into close proximity to each other, the F+ cell attaches to the F- cell by means of a sex pilus. Almost immediately, an F particle is transferred from the F+ cell to the F- cell. A few F- cells will incorporate the F particle into their chromosome to become Hfr (high frequency recombination) cells. This integration of the F particle appears to mobilize the chromosome for transfer to a recipient cell. As Hfr cells and F- cells come into contact, the donor chromosome breaks at the integration site of the F particle. The chromosome is not transferred as a circle, but as a linear sequence of genes. The gene that was adjacent to the F particle enters first and the origin and the F particle always enters last.

Lysogeny occurs when the DNA of a bacterial virus is incorporated into the DNA of a bacterial cell (Figure 1.3). During lysogeny, a bacteriophage (bacterial virus) attaches to a specific host cell. The phage DNA then penetrates the bacterial cell. The phage DNA then incorporates into the bacterial chromosome forming a prophage. The prophage is then replicated along with the bacterial chromosome. Occasionally, the prophage leaves the host cell DNA and begins to replicate whole phages. Thus the release of infectious phages is the best indicator that the cells of lysogenic.

Transduction occurs when bacterial DNA is carried within a bacteriophage from a donor bacterium to a recipient bacterium (Figure 1.4). The DNA of a phage enters a bacterial host cell. The phage DNA then

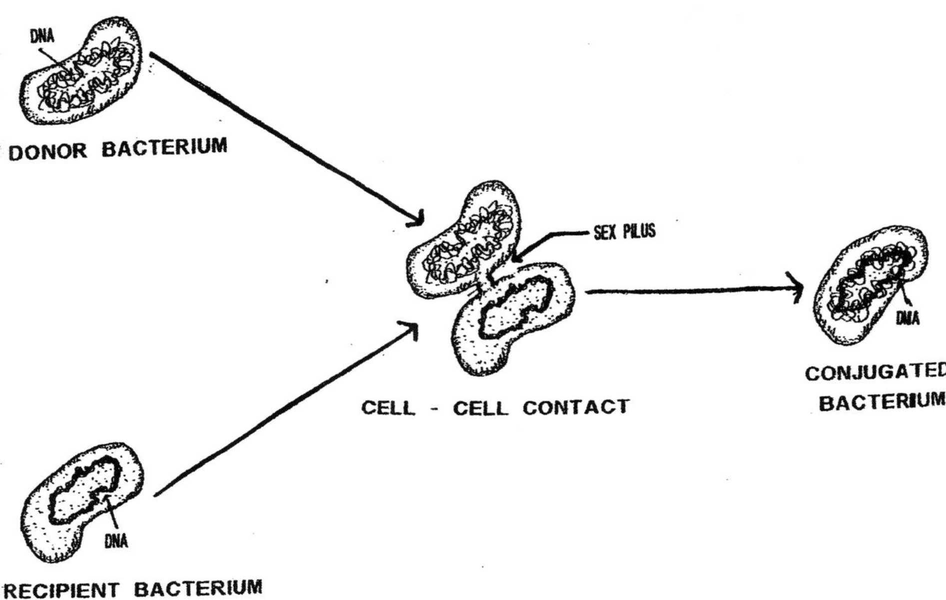

Figure 1.2. Conjugation. Two bacteria contact each other using a sex pilus. DNA is transferred from the donor bacterium into the recipient bacterium.

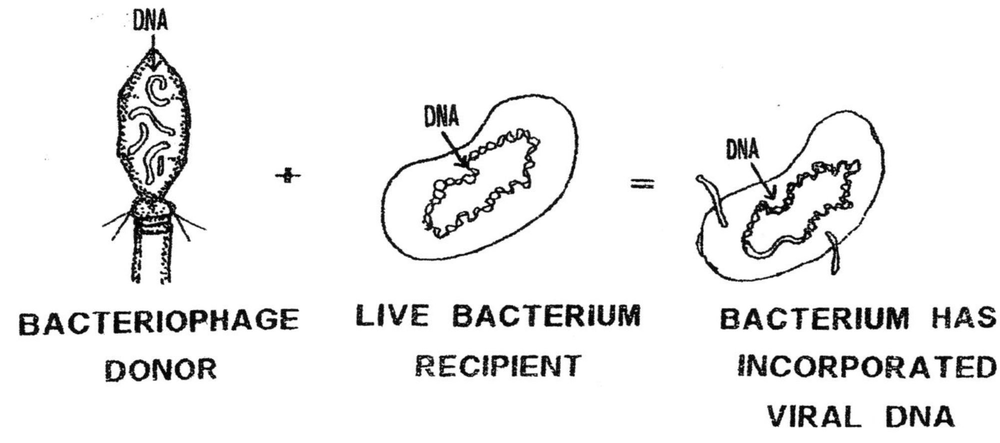

Figure 1.3. Lysogeny. DNA from a bacteriophage is incorporated into the DNA of a live bacterial recipient

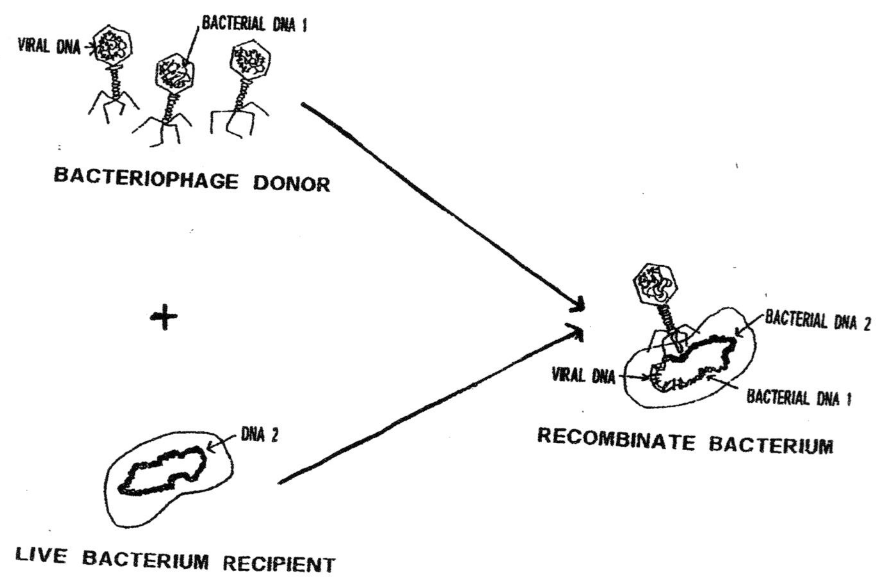

Figure 1.4. Transduction. Bacterial DNA is carried by a bacteriophage from a donor bacterium to a recipient bacterium.

becomes integrated with the host cell DNA forming a prophage. The phages that are replicated are considered defective in that DNA of the host bacterial cell has replaced some of the viral genes. When the defective phage DNA penetrates a new host bacterium, it cannot cause the replication of new phage particles. The bacterial genes introduced into the new host cell by the defective phage are integrated into the DNA of the new host bacterium.

A plasmid is genetic material that never integrates into the bacterial chromosome. The exchange of plasmids between to bacterial cells is referred to as **plasmid transfer** (Figure 1.5). This is an important mechanism in the transfer of antibiotic resistance between different strains of bacteria.

<u>The Evolution of Transgenic Sex</u>

Transgenic sex appears to have evolved from repair mechanisms of early DNA replicators. Prior to the onset of life, individual-replicating molecules of DNA frequently broke and reattached creating new variation and properties that increased the chance of molecular replication. Bombardment of the earth's surface with harmful solar ultraviolet radiation damaged the base sequence of these DNA molecules thus preventing replication. If the damage was to one strand of the DNA molecule, excision repair could be used to repair the damage (Figure 1.6). If damage occurred in the same place on both strands of the DNA replicator, a second DNA molecule was required to provide the base sequences to repair both strands of the damaged DNA molecule.

SEX AND REPRODUCTION

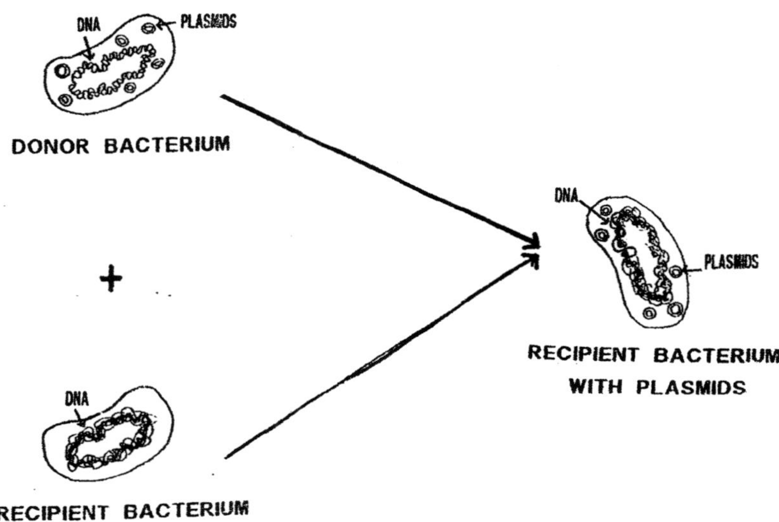

Figure 1.5. Plasmid transfer. Plasmids are transferred from a donor bacterium to a recipient bacterium.

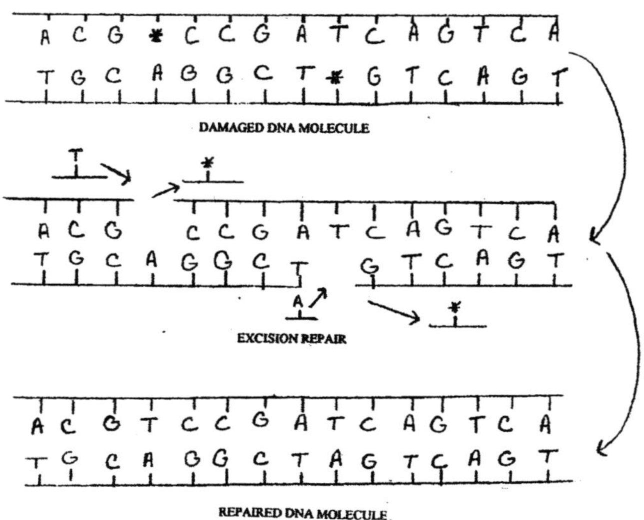

Figure 1.6. Repair among primitive DNA replicators. Damaged bases in a DNA stand are excised and replaced.

SEX AND REPRODUCTION

DNA replicators eventually became enclosed within plasma membranes giving rise to the first prokaryotic cells. Destruction of DNA by sunlight selected for genetic repair mechanisms in the early prokaryotes. In the process of repairing a single DNA strand, bacteria use nuclease enzymes to excise and remove the damaged portion of the DNA strand. Free nucleotides were then used to fill in the gap using the damaged uncomplimentary strand of DNA as a template. If the damage occurred in the same place on both strands of DNA, bacteria would have no template copy for repairing the DNA molecule. It is believed that bacteria started to use their repair enzymes to incorporate DNA of other bacteria (transgenic sex) to repair their damaged DNA molecule. When this started, sex as recombination had begun. Genes for this type of repair mechanism spread rapidly throughout the bacterial world. Sex as a genetic infusion permitted bacteria to survive a chemically erratic world.

CELLULAR SYMBIOGENESIS

<u>The Evolution of Cellular Symbiogenesis</u>

 Cellular symbiogenesis is the permanent merging of two or more cells to produce an entirely new cellular entity. Cellular symbiogenesis between prokaryotic cells occurred approximately 1.7 to 2 billion years ago in response to environmental stresses. Cellular symbiogenesis allowed the development of the first eukaryotic cells, which eventually evolved into the protoctists, fungi, animals, and plants.

In environmental conditions devoid of oxygen, the first symbiotic relationship appears to have evolved from the invasion of a thermoplasma-like archaebacteria by spirochetes. The symbiotic relationship that developed allowed the spirochete to have a constant stream of energy and food and gave the thermoplasma-like archaebacterium mobility to seek food resources, as well as avoid harsh environmental conditions and predators. As the new symbiotic entity developed, the acquisition of the centroile-kinetechore complexes from its spirochete ancestry allowed some early protoctists to replicate by mitosis.

In an acidic, hot, watery environment surrounded with atmospheric oxygen, a second symbiotic relationship appears to have occurred between mobile thermoplasma-like organisms mention above and purple bacteria capable of cellular respiration. The symbiotic purple bacteria that developed into mitochondria gave this new cellular entity the ability to survive an environment rich in oxygen, whereas the mobile symbiotic thermoplasma-like microorganism protected the purple bacterium from the acid, hot environment. Purple bacteria have been implicated in this major evolutionary event because they are similar in size to mitochondria. Mitochondria also reveal their prokaryotic heritage by replicating at times different from their host. Lastly, the DNA nucleotide sequence of modern purple bacteria is similar to that observed for mitochondria of present day eukaryotic cells.

The most recent symbiotic relationship appears to have taken place between an early protoctist and a coccoid cyanobacterium capable of photosynthesis. The eukaryotic cellular entity that formed eventually evolved into the various plants that we see today. Like mitochondria, plastids of modern plants reveal their prokaryotic heritage by replicating at times different from their host cells. Plastids in modern plants are also similar in size and shape to present-day coccoid cyanobacteria. Finally, the nucleotide sequence of plastids of modern plants is similar to that found in present day coccoid cyanobacteria, suggesting a prokaryotic heritage.

The evolution from symbiont to organelle involved the transfer of many genes from the symbiont to the host cell nucleus. The mitochondrial genome of modern day eukaryotic cells only contains around a dozen genes. This is not sufficient to maintain the survival of isolated mitochondria *in vitro* for more than several hours. The genome of modern day plastids contains only between 100 genes to 200 genes. Since most of the genome of mitochondria and plastids have been transferred to the nucleus, it is important that proteins encoded by those genes be transferred back to their appropriate organelle. This is accomplished by a string of amino acids at the end of the translated protein that is recognized by the membrane of the organelle, which facilitates its transfer into that organelle.

MEIOTIC SEX

The Advantages of Meiotic Sex

Meiotic sex (also known as fertilization sex) is the life-cycle combination of meiosis and fertilization that is found in sexual eukaryotic organisms. Meiotic sex appears to have evolved in some of the descendents of eukaryotic cells that had evolved from cellular symbiogenesis. In meiotic sex, haploid propagules fuse to form a diploid state, that in turn forms haploid propagules.

Reproduction involving meiotic sex with recombination has numerous advantages over asexual reproduction. Organisms practicing meiotic sex with recombination produce progeny with genetic variability that can survive different or rapidly changing environments. Organisms practicing meiotic sex can easily repair damaged DNA. Sexual populations of organisms evolve more rapidly and accumulate considerably less deleterious mutations when compared to their asexual counterparts.

The Evolution of Meiotic Sex

Meiotic sex appears to have first evolved in some haploid protoctists exposed to alternating periods of optimal and harsh environmental conditions (e.g. cold and drought). During optimal environmental conditions, these haploid protoctists grew and multiplied. To survive seasonal harsh environmental conditions, these early protoctists may have been forced to become diploid, only to return to a haploid state by meiosis when optimal conditions returned. There are several possible ways in which diploidy may have occurred. First, some early protoctists may have been forced to practice cannibalism. This would result in the merging of all cellular contents including the genetic material. Diploidy may also have occurred by **endomitosis** without cell division (Figure 1.7). This can be observed today in several modern life forms. When optimal environmental conditions returned, **one-step meiosis** allowed these early protoctists to return to a haploid state to resume maximal growth and reproduction.

As early meiotic sex of protoctists evolved, haploid protoctists began to create the diploid state by **cellular fusion** (Figure 1.8). Two haploid protoctists suffering from double stranded DNA damage would most likely have damage in different areas of the DNA molecules. Creation of a diploid cell by cellular fusion would allow for the repair of both DNA molecules before the creation of the haploid condition by one step-meiosis.

With the advent of diploidy through cellular fusion, one-step meiosis was eventually replaced with **two-step meiosis with genetic recombination** (Figure 1.9). Cells replicating using two-step meiosis with genetic recombination had several advantages. Chromosomal recombination during diploidy enhanced the rate of evolution. Chromosomal recombination also lowered the expression of deleterious mutations.

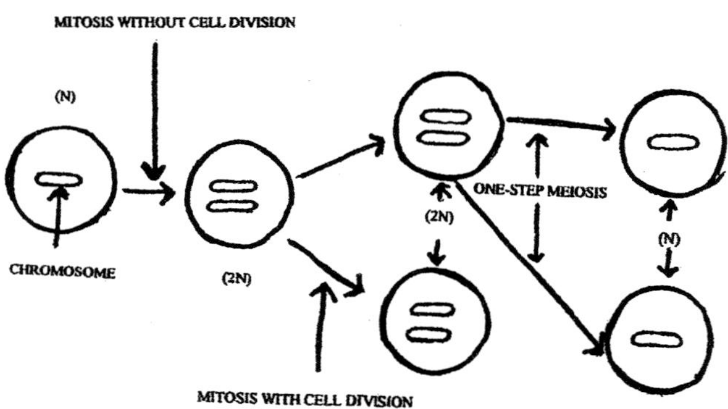

Figure 1.7. Diploidy by endomitosis. A cell becomes diploid without cell division. Cells return to a haploid state by one-step meiosis. Drawing by Mark S. Kosco.

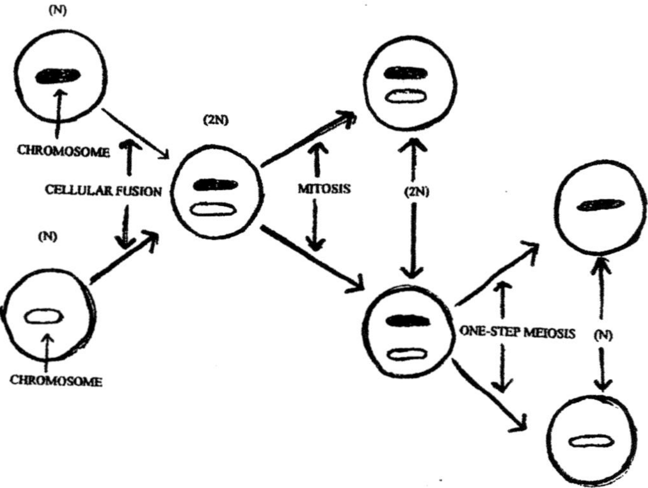

Figure 1.8. Diploidy by cellular fusion. Two haploid protoctists fuse to form a diploid cell. Diploid protoctists return to a haploid state by one-step meiosis.

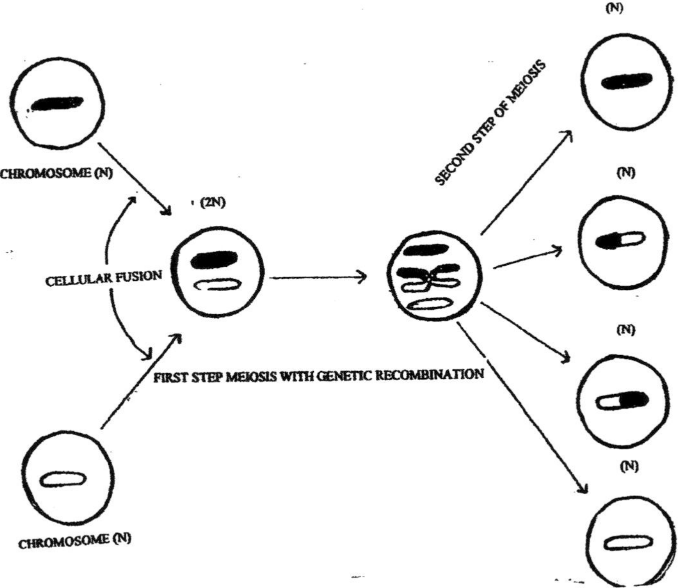

Figure 1.9. Two-step meiosis with genetic recombination. With the advent of diploidy by cellular fusion, two-step meiosis allowed genetic recombination to take place that increased the genetic variability when protoctists returned to a haploid state.

<u>Diploidy in the Evolution of Multicellularity</u>

Multicellularity and increasing complexity has been the hallmark of evolution for the last 550 million years. As animal and plant life evolved, more genes were required to build organisms as compared to their unicellular ancestors. With an increase in the number of genes there was also an increase in the amount of DNA. With a fixed error rate per replicating nucleotide, an increase in the amount of DNA to be replicated would mean more mutations in life forms with large genomes. Therefore, mechanisms needed to be found to eliminate or reduce the number of mutations when copying an increasing number of nucleotides.

As more complex forms of single celled organisms evolved, there was an increase in the number of genes. To prevent an increase in the number of mutations in the genome, these organisms increase the efficiency in which the nucleotides were replicated. However as life became multicellular, and the genome increased in size, the mutation rate per genome dramatically increased. To control the mutation rate per genome, multicellular life opted for diploidy.

In the first primitive multicellular organism, being diploid for most of the life cycle had advantages over being haploid. The majority of deleterious mutations is recessive and could be masked in the diploid state. Since recessive mutations accumulate in the genome and begin to express themselves as their numbers increase, outcrossing and meiotic sex with recombination became a valuable part of the lifecycle of complex multicellular organisms. Outcrossing and recombination followed by two-step meiosis became important in regulating the mutational load of the genome. Outcrossing, recombination and two-step meiosis also provided a mechanism for repairing damaged DNA in the haploid propagules. Fusion of haploid propagules during fertilization would then create a new organism essentially free of genetic damage.

2

THE NEUROENDOCRINOLOGY OF MAMMALIAN REPRODUCTION

INTRODUCTION

Hormones are chemical substances that are produced by endocrine glands and are transported to their target organs via the blood. Hormones can be chemically divided into steroids, glycoproteins, and amines. Steroid hormones have a cyclopentanoperhydrophenanthrene ring system of sterols and are all derivatives of cholesterol. **Testosterone, estrogen** and **progesterone** are common steroid hormones. Glycoprotein hormones upon decomposition yield proteins and carbohydrates. Examples of glycoprotein hormones are **follicle stimulating hormone (FSH), luteinizing hormone (LH)** and **chorionic gonadotropin (CG)**. Amines are any member of a group of compounds formed by replacing one or more of the hydrogens of ammonia by one or more univalent hydrocarbons or other nonacid organic radicals. Examples of amines include **epinephrine** and **norepinephrine**.

MAJOR HYPOTHALAMIC HORMONES INVOLVED IN MAMMALIAN REPRODUCTION

Corticotropin Releasing Hormone

Human **corticotropin releasing hormone (CRH)** is a 41 amino acid polypeptide that is identical to that found in the rat. Human CRH is synthesized from a precursor of 196 amino acids and stimulates the release of **adrenocorticotropic hormone** (ACTH) and other products of **pro-melanocortin**. The half-life of human CRH is approximately 60 minutes. The parvocellular neurons that secrete CRH are found in the anterior portion of the paraventricular nuclei just lateral to neurons that secrete **thyrotropin releasing hormone (TRH)** (Figure 2.1). There is a significant increase in the level of plasma CRH in the human female during late pregnancy and parturition. A specific CRH binding protein has recently been discovered in the blood plasma and intracellular fluids of cells.

Gonadotropin Releasing Hormone

Gonadotropin releasing hormone (GnRH) is a linear peptide composed of 10 amino acids. Pro-GnRH, a precursor of GnRH, is composed of 92 amino acids. Pro-GnRH has a 23 amino acid sequence, a 10 amino acid peptide (GnRH), a 3 amino acid proteolytic processing site and a 56 amino acid sequence called GnRH-associated peptide (GAP). GAP has existed for 500 million years and is structurally identical in all mammals. GAP is present in mammalian neural and non-neural tissues and stimulates FSH and LH secretion, and experimentally inhibits the secretion of prolactin. The physiological role of GAP is not well understood.

Gonadotropic releasing hormone is present in different mammalian and non-mammalian species. The structure of GnRH varies slightly among different mammalian species. Neurons that secrete GnRH are primarily located in the preoptic area of the anterior hypothalamus (Figure 2.1) and their axons are found in the lateral portions of the external layer of the median eminence adjacent to the pituitary stalk.

THE NEUROENDOCRINOLOGY OF MAMMALIAN REPRODUCTION

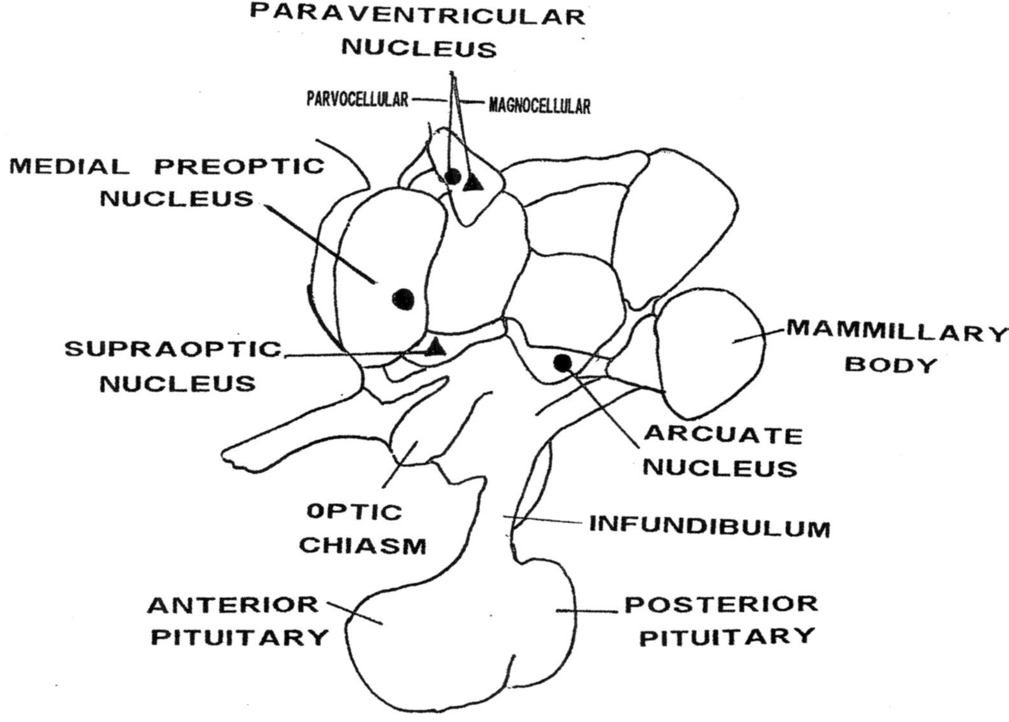

Figure 2.1. Hypophysiotropic neurons involved in mammalian reproduction. A saggital view of the hypothalamus and pituitary showing the location of hypophysiotropic neurons.

Prolactin Releasing Factors

Prolactin releasing factors (PRL releasing factors) are a group of chemicals that stimulate the release of prolactin from the mammalian anterior pituitary. **Thyrotropin releasing hormone (TRH)** is the best studied of the PRL releasing factors. Thyrotropin releasing factor is a tripeptide that is produced from a 242 amino acid precursor that contains six copies of TRH. Neuron cell bodies that secrete TRH are located in the medial portions of the paraventricular nuclei with their axons terminating in the medial portion of the external layer of the median eminence. **Vasoactive intestinal peptide (VIP)** is a peptide composed of 28 amino acids that is also produced by the hypothalamus. Vasoactive intestinal peptide stimulates the secretion of prolactin in humans. Serotonin precursors administered to humans have shown to stimulate the release of prolactin from the anterior pituitary.

Prolactin Inhibiting Hormone

Prolactin inhibiting hormone (PIH) is a mammalian hypothalamic secretion that inhibits the release of prolactin from the anterior pituitary. **Dopamine** is the primary prolactin inhibiting hormone. Dopamine is an intermediate of tyrosine metabolism and a precursor of **norepinephrine** and **epinephrine**. The cell bodies of dopamine secreting neurons are located in the arcuate nuclei with their axons terminating in the external layer of the median eminence (Figure 2.1). **Gamma-aminobutyric acid (GABA)** and cholinergic pathways have also showed to inhibit prolactin secretion in some mammalian species.

11

THE NEUROENDOCRINOLOGY OF MAMMALIAN REPRODUCTION

<u>Oxytocin</u>

Oxytocin is a small peptide of 9 amino acids characterized by a ring structure with a disulfide bond. In mammals, the prohormone for oxytocin is synthesized by magnocellular neuron cell bodies located in the supraoptic nuclei and the lateral and superior parts of the paraventricular nuclei. The prohormone form of oxytocin is transported via the hypothalamo-hypophyseal tract to the posterior pituitary. In the axon terminals, the prohormone (contained in secretory granules) is processed producing oxytocin and **neurophysin I**. Neurophysin I serves as a transport protein for oxytocin.

Oxytocin generally causes contraction of smooth muscle tissue. It is involved in uterine contractions during parturition, contraction of myoepithelial cells surrounding the alveoli of the mammary gland during milk-let down and contractions of the male reproductive tract during ejaculation.

MAJOR PITUITARY HORMONES INVOLVED IN MAMMALIAN REPRODUCTION

<u>The Structure of Gonadotropins</u>

A **gonadotropin** is a gonad-stimulating hormone. The principal source of gonadotropins is the anterior pituitary gland (e.g. FSH and LH), the urine of pregnant animals (e.g. hCG) and the serum of pregnant mares (e.g. eCG).

All gonadotropins are composed of two dissimilar units referred to as alpha and beta.   The primary structure of the alpha subunits among the various gonadotropins is similar but not identical. The biological specificity of gonadotropic hormones can be found in the primary structure of the beta subunit. Hybrid combinations of the alpha subunit of one glycoprotein hormone with the beta subunit of another result in the recovery of considerable biological activity of the hormone from which the beta subunit was obtained.

The carbohydrate content of gonadotropins of various mammalian species ranges from 13% to 45%. Within these glycoproteins the monosacchrides mannose, galactose, fucose, N-acetylglucoseamine, N-acetylgalactosamine and sialic acid are linked in multiple oligiosacchride chains to form complex heteropolysacchride units. Sialic acid is essential for the full expression of the biological activity of several gonadotropic hormones and prolongs their half-lives by inhibiting their uptake and degradation by the liver.

Gonadotropic hormones are polymorphic between and within mammalian species. The polymorphism c*
generally be attributed to the microheterogeneity of the carbohydrate units, deletion of amino acids from
the amine and carboxyl end of the polypeptide chains, and differences in the sialic acid content.

Luteinizing hormone (LH) in most mammalian species has a molecular weight of approximately 3C
but is slightly larger in primates and horses. Sialic acid content varies between species and is foun*
high levels in equine LH. The primary structure of the alpha subunit varies between 89 – 96 amin*
most mammalian species. The beta subunit of mammalian LH contains approximately 119 amin*
LH has several functions in the male and female. A surge in plasma LH has been shown to caus*
in numerous mammalian species. Luteinizing hormone is also involved in the initiation of lutei*
is responsible for the maintenance of luteal function. Luteinizing hormone plays a dominant r*

dominant role in steroid synthesis in the male as indicated by the expansion of the Golgi complex (Camillo Golgi, Italian histologist, 1843 – 1926) and endoplasmic reticulum of Leydig cells (Franz von Leydig, German anatomist, 1821 – 1908).

Follicle stimulating hormone (FSH) is a glycoprotein with a molecular weight of approximately 32,000. The alpha subunit contains approximately 89 amino acids, whereas the beta subunit contains around 115 amino acids. FSH has several physiological functions in the male and the female. Follicle stimulating hormone is needed for follicular development in the female.   A small surge in plasma FSH can be observed prior to ovulation and may be responsible for stimulating the growth and maturation of future generations of follicles. FSH may also be involved in the initiation of luteinization in some species of mammals, although this is the primary function of LH. Follicle stimulating hormone is required for spermiogenesis in the male.   Follicle stimulating hormone also stimulates the Sertoli cells (Enrico Sertoli, Italian histologist, 1842 – 1910) of the testes to produce and secrete a multitude of proteins that include the likes of androgen binding protein and inhibin.

Chorionic gonadotropin (CG) is a hormone that was first discovered in the urine of pregnant mammals by Ascheim and Zondek [Selmar Ascheim and Bernhard Zondek, Israeli gynecologists and endocrinologists] in 1927. **Human chorionic gonadotropin (hCG)** is produced by the placenta and reaches maximum urinary concentrations between the 10th and 12th week of pregnancy, but declines to lower levels during the last half of pregnancy.   Human chorionic gonadotropin is slightly larger than FSH and LH with a molecular weight around 40,000. The alpha subunit contains about 92 amino acids and its primary structure is identical to LH except with additional 3 amino acids at the amine terminal. The primary structure of the beta subunit of hCG is similar to the chorionic gonadotropins found in sheep and pigs. The biological properties of hCG are similar to that of LH. Human chorionic gonadotropin stimulates the theca cells of the ovary to produce estrogens, causes luteinization of the granulosa cells and maintains function of the corpus luteum.   The administration of hCG to males causes the Leydig cells to produce androgens.

Prolactin

Human **prolactin (PRL)** is a single chain polypeptide of approximately 198 amino acids and has 3 intrachain disulfide bonds. Prolactin from sheep, pigs and humans has a molecular weight around 22,500. The primary structure of PRL has been found to be similar to that for growth hormone.

Autoradiographic studies have shown that prolactin receptors are located on the Leydig cells of the testes. Numerous studies have shown that prolactin appears to be responsible for regulating the yield of spermatozoa produced by the seminiferous tubules. The exact mechanism of action remains unclear. Prolactin has also been shown to work synergistically with LH to stimulate testicular androgen secretion. Prolactin has been reported to constitute the luteotropic complex required for the maintenance of the corpus luteum in several mammalian species. The luteotropic complex consists of prolactin and LH; however, in the hamster prolactin and FSH are required. Prolactin is also involved in the development of the duct system of the mammary gland during pregnancy and has also been shown to suppress the ovarian cycle during lactation.

THE NEUROENDOCRINOLOGY OF MAMMALIAN REPRODUCTION

MAJOR GONADAL AND NONGONADAL FACTORS INVOLVED IN MAMMALIAN REPRODUCTION

Nonsteroidal Gonadal Regulatory Hormones

The **inhibins** are nonsteroidal gonadal factors that are members of the transforming growth factor-B family (Figure 2.2). The inhibins are dimers of alpha and beta subunits linked by disulfide bonds. Inhibins exist in two heterodimeric forms. **Inhibin-A** consists of an alpha subunit and a beta$_A$ subunit. **Inhibin-B** consists of an alpha subunit and a beta$_B$ subunit. Sertoli cells of the testes and granulosa cells of the ovarian follicle secrete inhibin. Inhibin selectively inhibits FSH secretion in both the male and the female.

The **activins** are nonsteroidal gonadal factors that are members of the transforming growth factor-B family (Figure 2.2). The activins are dimers of beta subunits of inhibin.   The activins exist in three forms. **Activin-A** is a homodimer of two beta$_A$ subunits. **Activin-B** is a homodimer of beta$_B$ subunits. **Activin** is a heterodimer of a beta$_A$ subunit and a beta$_B$ subunit. Activins augment FSH secretion but inhibit the secretion of GH, ACTH and PRL in the female. Activin also appears to enhance the pituitary response to GnRH.

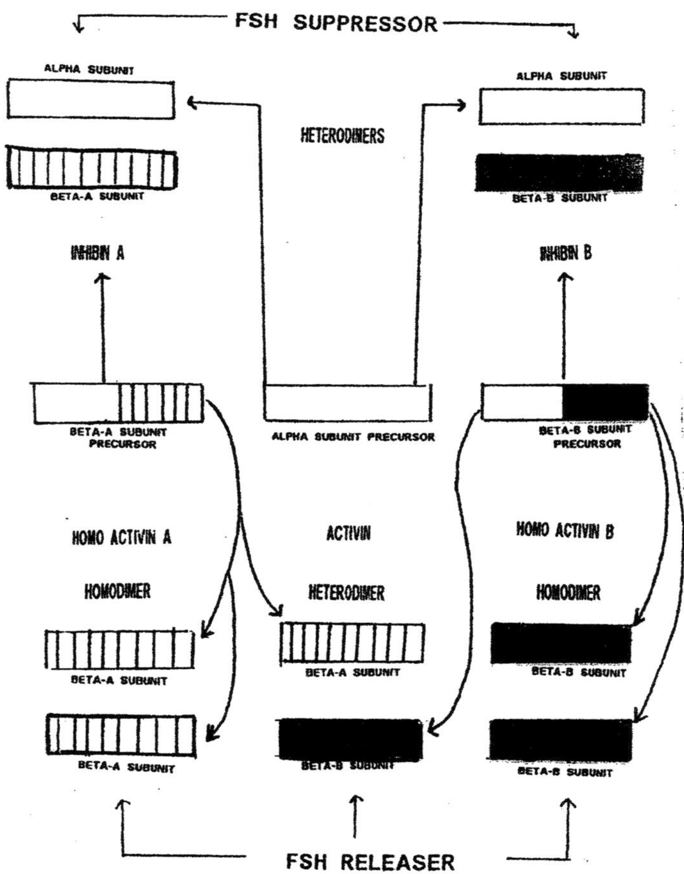

Figure 2.2. Structure of the inhibins and the activins.

THE NEUROENDOCRINOLOGY OF MAMMALIAN REPRODUCTION

Mammalian **follistatin** is a single chain glycoprotein that is produced by the ovary and a variety of pituitary cells. Follistatin inhibits FSH synthesis and secretion as well as pituitary's response to GnRH stimulation. The probable mechanism of action for follistatin is the binding of activin. Follistatin may act as an intratesticular paracrine agent regulating spermatogenic and interstitial function.

Pro-opiomelanocortin (POMC) is a molecule that is found in many tissues of the mammalian body including the pars intermedia of the adenohypophysis, hypothalamus, sympathetic nervous system, gonads, placenta, gastrointestinal tract and lungs. Pro-opiomelanocortin serves as a precursor protein for numerous other substances including ACTH, corticotropin-like intermediate lobe protein (CLIP), beta-lipotropic hormone (B-LPH), gamma melanocyte stimulating hormone (gamma-MSH), beta–endorphin and met-enkephalin. Some of these opioids influence hypothalamic functions involved in reproduction. These opioids will be discussed later.

MAJOR GONADAL STEROIDS INVOLVED IN MAMMALIAN REPRODUCTION

The Structure of Steroid Hormones

Steroids are a large group of compounds that have a general structural resemblance to terpenes. Members of this chemical group consists of three fused cyclohexane rings (rings A, B and C) and a terminal cyclopentane ring (ring D). Most naturally occurring steroids have an oxygenated substitute on the third carbon and angular methyl groups (numbered 19 and 18) on carbons 10 and 13. The most abundant sterol in animals is cholesterol, and this has been found in almost every tissue of the body. Cholesterol is synthesized from acetate primarily in the smooth endoplasmic reticulum of the cell. Cholesterol is transported to the mitochondrion for cleavage of the side chain between carbons 20 and 22 resulting in the formation of pregnenolone. Progesterone and other steroids can be synthesized from pregnenolone (Figure 2.3).

Pregnenolone serves as a precursor for the synthesis of androgens (male sex hormones). **Androgens** are C19 steroids and belong to a class of steroids devoid of the carbon side chain at position 17. **Testosterone** is produced in the testes and is a key androgen in the male. The ketone group at position 3 as well as the hydroxyl group at position 17 is important for testosterone's biological activity. Other androgens include **androstenedione** and **dihydrotestosterone** that are produced to a lesser amount by the testes. The **estrogens** are derived from the androgens by elimination of the C19 methyl group and aromatization of the A ring. Consequently, there is no available valence for the methyl group at position 10, and the hydroxyl group at position 3 is phenolic instead of alcoholic. The rings and the substitutes on C17 determine the biological activities of the estrogens. The three major estrogens are **estrone, estriol** and **estradiol-17B**. Estradiol-17B is approximately 12 times more potent than estrone, while estrone is approximately 80 times more potent than estriol in the human female.

Physiological Effects of Androgens in the Male

In many mammalian species there are three periods of elevated plasma testosterone concentration during the life of the male (Figure 2.4). During human fetal development, there are two periods of elevated testosterone secretion by the testes. The first period occurs at approximately the seventh week of gestation.

THE NEUROENDOCRINOLOGY OF MAMMALIAN REPRODUCTION

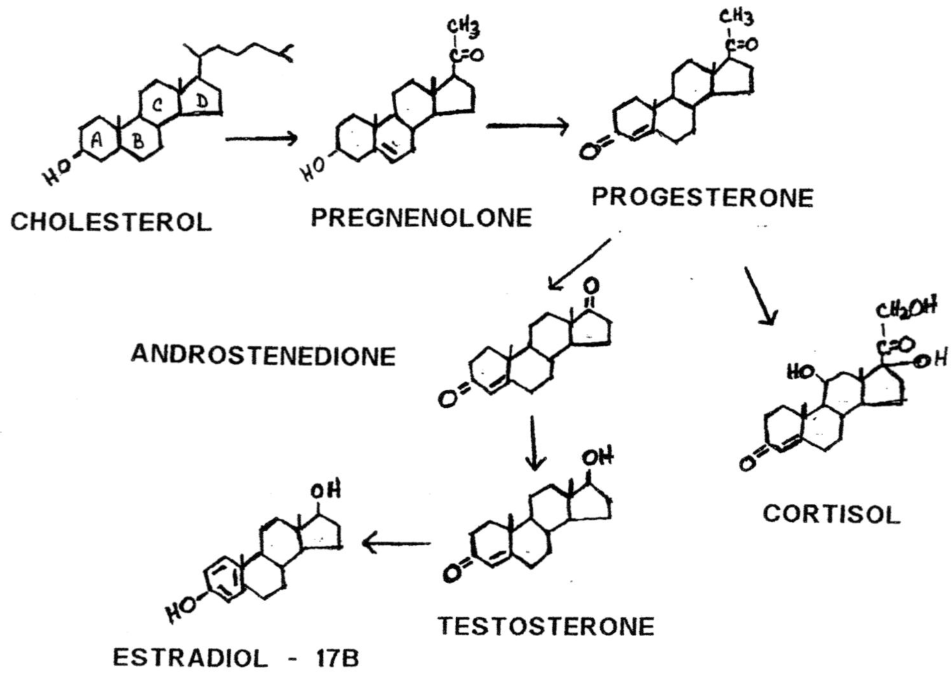

Figure 2.3. The synthesis of major steroid hormones involved in mammalian reproduction.

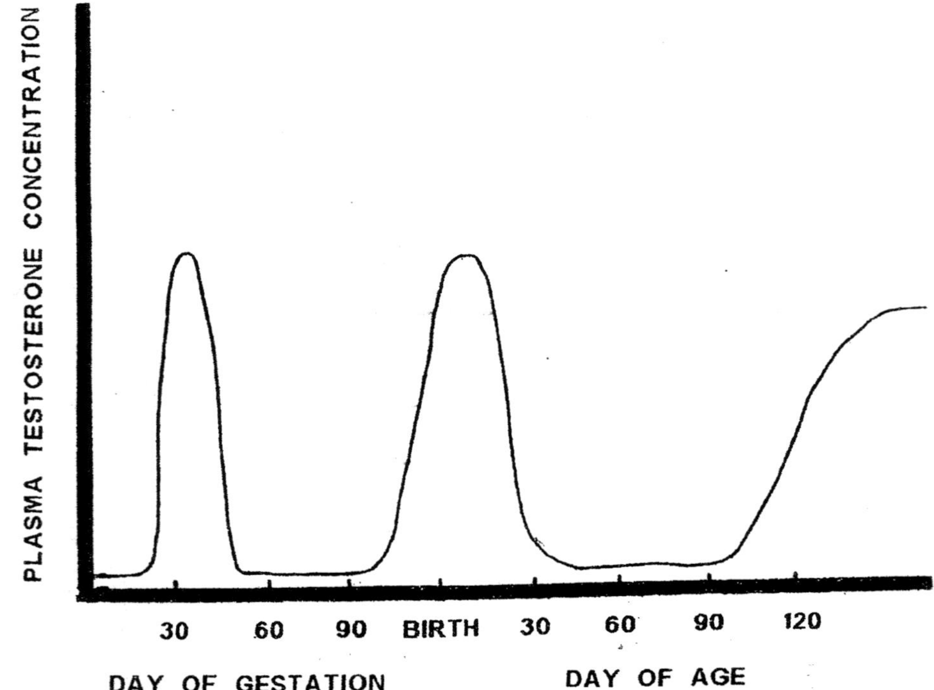

Figure 2.4. Plasma levels of testosterone in the boar from early gestation through puberty.

THE NEUROENDOCRINOLOGY OF MAMMALIAN REPRODUCTION

This is responsible for the development of the accessory sex glands. The second period occurs several weeks prior to and after birth. During this time, testosterone is responsible for the descent of the testes into the scrotum. The last period of elevated plasma testosterone occurs at the time of puberty. During this time, testosterone has a wide variety of effects on the male that include increased skeletal development, hypertrophy of the larynx, decreased growth of hair on top of the head, increased metabolic rate, increased erythrocyte production, and enlargement of the penis, scrotum and testes. Similar effects are seen in other species of animals.

<u>Physiological Effects of Estrogens in the Female</u>

Estrogens are found in minute quantities during female prepubertal development in several mammalian species. However, the secretion of estrogens by the ovary increases dramatically at the time of puberty. Elevated pubertal levels of estrogens have a dramatic influence on the female reproductive tract and mammary gland. There is a marked increase in the size of the ovaries, oviducts and uterus. There is an increase in the number of ciliated epithelial cells as well as an enhancement of ciliary activity in the oviduct. Proliferation of stromal tissue and increased development of endometrial glands occurs in the uterus. The external genitalia enlarge due to increased fat deposition in the labia minora, labia majora and mons pubis. Estrogens cause an increase in stromal development, growth of the duct system and fat deposition in the mammary gland. There is also a slight increase in lobule and alveolar development.

Elevated pubertal estrogen levels have an influence on bone development and metabolism in human females. Estrogens cause an increase in osteoblast activity resulting in an increase in early pubertal growth. Estrogens are also responsible for the early uniting of the epiphysis with the shaft in long bones. This causes an earlier cessation of pubertal growth compared to the male. Estrogens cause an increase in metabolic rate resulting in a slight increase in total body protein and subcutaneous fat. Estrogens are also responsible for the increase in cutaneous vascularization.

<u>Physiological Effects of Progesterone in the Female</u>

Progesterone has a wide variety of effects on the reproductive tract of pubertal and post pubertal mammalian females. Elevation of progesterone levels during the second half of the menstrual cycle and estrual cycle promotes glandular secretion of the uterine endometrium and secretory changes in the mucosal lining of the oviduct. Progesterone is also responsible for the development of the lobules and cellular proliferation within the alveoli of the mammary gland. Progesterone also promotes an increase in sodium, chlorine and water retention by the kidneys.

CELLULAR MECHANISM OF ACTION FOR STEROID AND NONSTERIOD HORMONES

<u>Cellular Mechanism of Action for Protein Hormones</u>

Protein hormone regulation of target organ cellular activity involves a secondary messenger system (Figure 2.5). The protein hormone attaches to a transmembrane protein receptor. In the presence of magnesium, guanosine triphosphate (GTP) is exchanged for guanosine diphosphate (GDP) on the membrane bound G protein complex. The activated alpha subunit of the G protein complex disassociates and attaches to membrane bound adenylate cyclase. Activated adenylate cyclase catalyzes the conversion of adenosine triphosphate (ATP) to cyclic adenosine monophosphate (c-AMP). Cyclic AMP initiates an

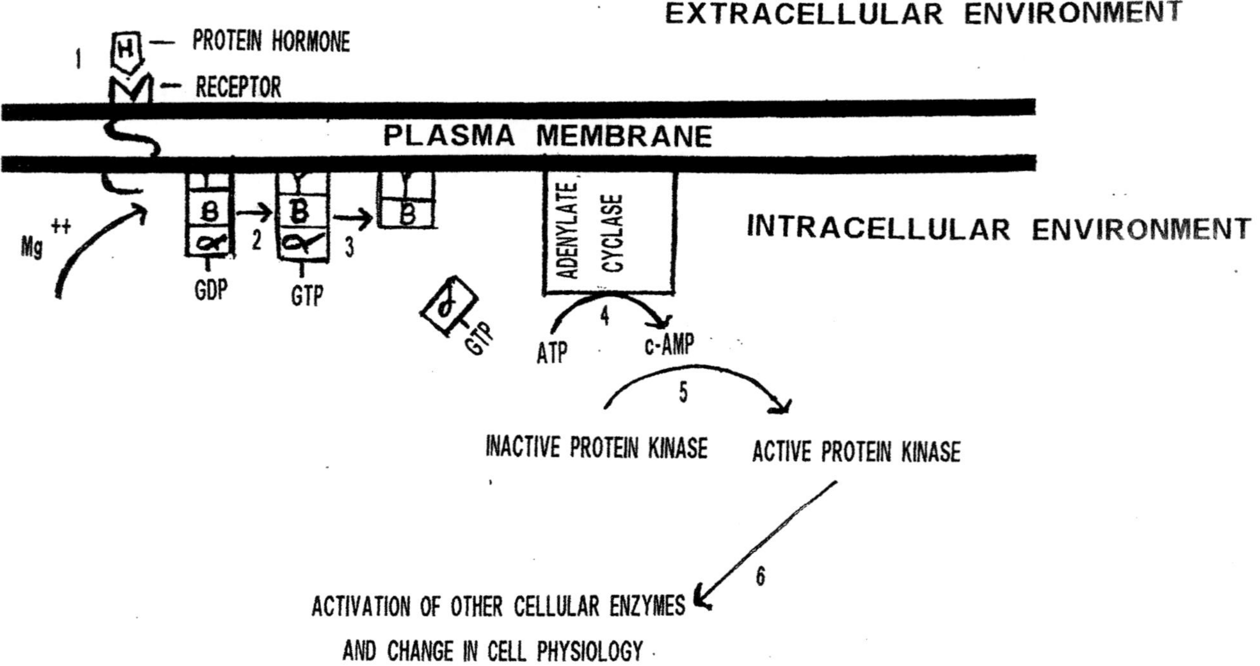

Figure 2.5. General mechanism for the action of nonsteroid substances at the cellular level. (1) Protein hormone binds to the membrane receptor. (2) Ionic magnesium aids in the exchange of GTP for GDP on the gamma subunit of the G protein. This activates the subunit of the G protein. (3) Alpha subunit of the G protein disassociates from the beta subunit and activates adenylate cyclase. (4) Adenylate cyclase catalyzes the conversion of ATP to c-AMP. (5) C-AMP initiates an enzyme cascade by activating protein kinase. (6) Protein kinase continues the enzyme cascade causing a change in the physiology of the cell.

enzyme cascade eventually resulting in a change in cellular function.

<u>Cellular Mechanism for the Action of Steroid Hormones</u>

The general mechanism of action for steroid hormones differs from that of proteins (Figure 2.6). Because steroid hormones are lipid soluble, steroid hormones easily penetrate the plasma membrane and nuclear membrane of their target cells to bind to specific nuclear receptors. The hormone-receptor complex then

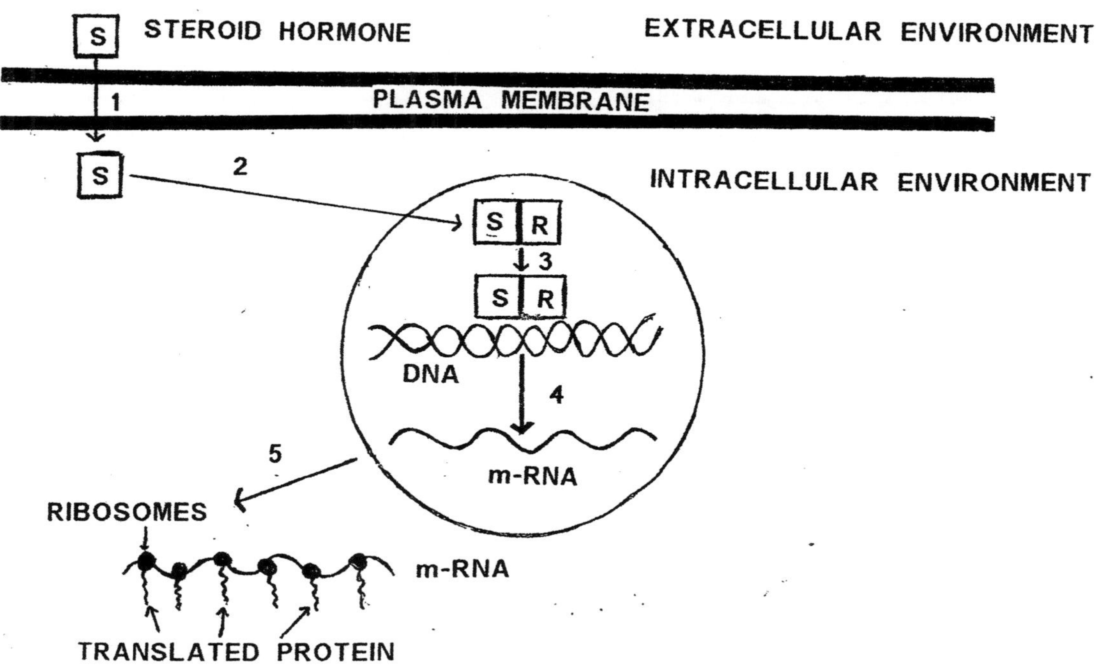

Figure 2.6. General mechanism of action of steroids at the cellular level. (1) Steroid penetrates the plasma membrane of the cell. (2) Steroid penetrates nuclear membrane and binds to a nuclear receptor. (3) Steroid-receptor complex binds to an acceptor site on the DNA. (4) Hormone-receptor complex initiates transcription resulting in the production of m-RNA. (5) m-RNA leaves the nucleus and is transcribed by cytoplasmic ribosomes resulting in a change in the physiology of the cell.

binds to specific points on the DNA strand. Transcribed m-RNA leaves the nucleus and diffuses into the cytoplasm where it is translated into protein by ribosomes. This protein alters cellular function. The mechanism of action for testosterone is slightly different from the general model of steroid action described above. When testosterone enters the cytoplasm of its target cell, it is converted into dihydrotestosterone. Dihydrotestosterone then binds to a cytoplasmic receptor. This hormone-cytoplasmic receptor complex then enters the nucleus resulting in transcription.

REGULATION OF PITUITARY HORMONE SECRETION

<u>The Hypothalamo-hypophyseal Portal System and the Hypothalamic-hypophyseal Tract</u>

The **hypothalamo-hypophyseal portal system** is a network of blood vessels that connect the hypothalamus to the anterior pituitary (Figure 2.7). This system begins with the primary capillary plexus in the median eminence of the hypothalamus. Blood contained in the primary capillary plexus then flows into portal vessels (small veins) that travel along the pituitary stalk to the secondary capillary plexus in the anterior pituitary. Hypothalamic releasing and inhibiting hormones that are secreted into the primary capillary plexus can be transported via the portal vessels to the secondary capillary plexus of the anterior pituitary.

The **hypothalamo-hypophyseal tract** (Figure 2.8) is nerve fibers that arise from cell bodies in the hypothalamus, travel down the infundibulum and end in the posterior pituitary. Hormones produced by the neuron cell bodies located within the hypothalamus travel to the posterior pituitary via their fiber tracts.

<u>Regulation of Anterior Pituitary Hormone Secretion</u>

For those anterior pituitary hormones whose production and secretion are dependent on hypothalamic releasing hormones, three types of negative feedback systems exists between the anterior pituitary and the

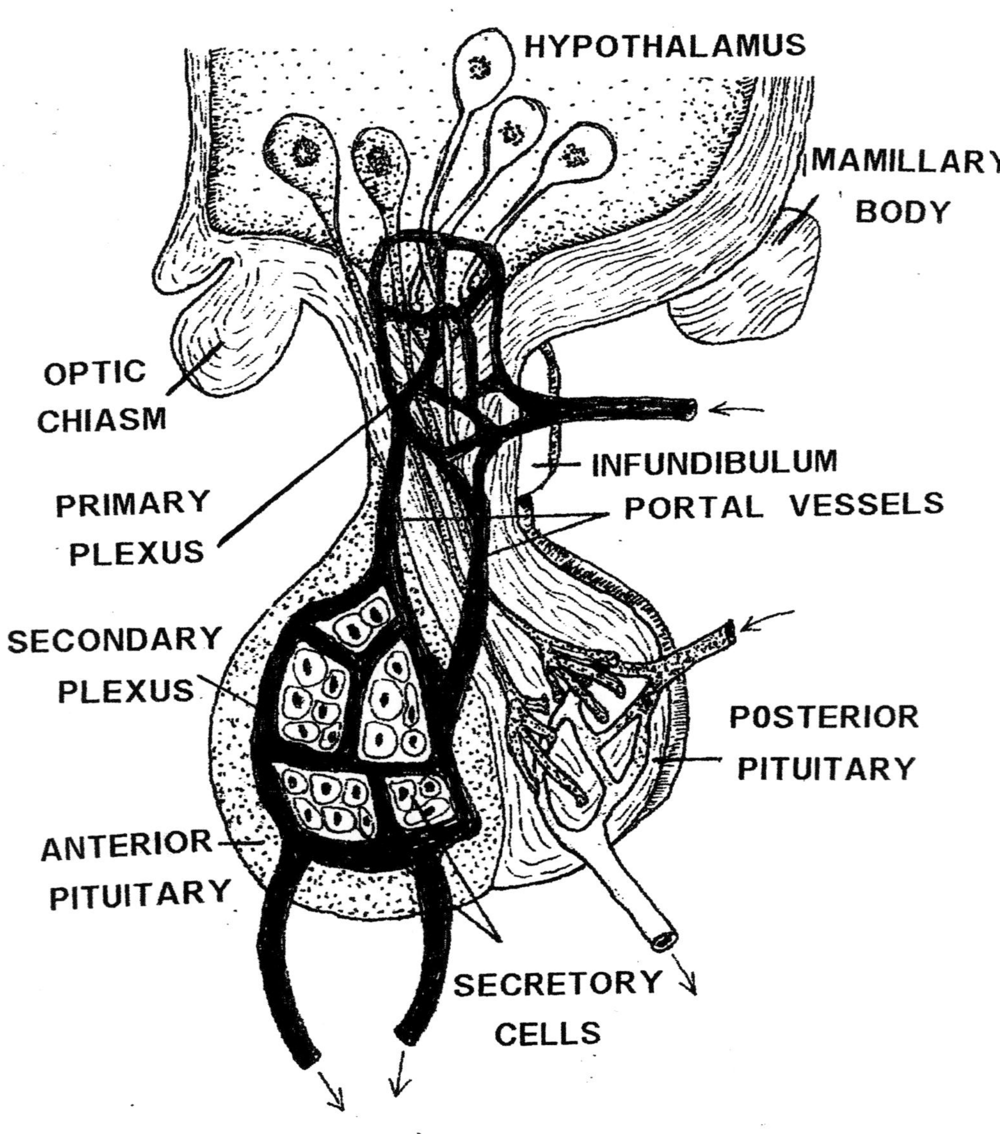

Figure 2.7. The hypothalamo-hypophyseal portal system.

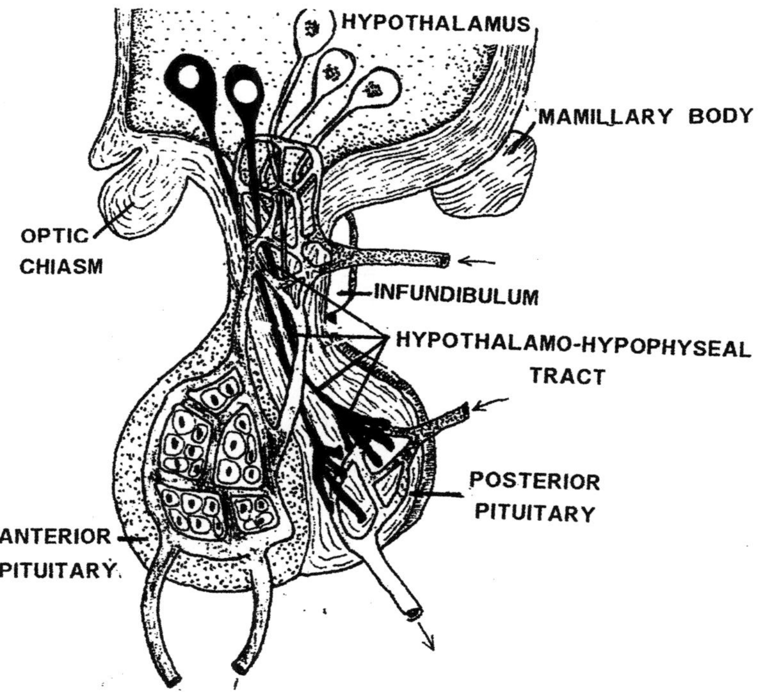

Figure 2.8. The hypothalamo-hypophyseal tract.

specific target organs (Figure 2.9). In **a long loop negative feedback system**, the anterior pituitary hormone acts on the target organ to increase the production of its hormone. The target organ hormone exerts a negative feedback on the hypothalamus and the anterior pituitary. In a **short loop negative feedback system** the anterior pituitary hormone exerts a negative feedback on the hypothalamus. In an **ultrashort loop negative feedback system** the hypothalamic-releasing hormone has a negative feedback on the hypothalamus regulating its own secretion. The negative feedback regulation of anterior pituitary hormones involved in mammalian reproduction will be discussed later.

For those anterior pituitary hormones whose synthesis and secretion is dependent on both a specific hypothalamic releasing hormone and a hypothalamic inhibiting hormone, the levels of these hypothalamic hormones often determines the plasma concentration of the anterior pituitary hormone. The regulation of secretion of these anterior pituitary hormones involved in mammalian reproduction will be discussed later.

<u>The Regulation of Posterior Pituitary Hormone Secretion</u>

Neuroendocrine reflexes regulate the secretion of hormones (e.g. oxytocin and antidiuretic hormone) from the posterior pituitary (Figure 2.10). The release of hormones from the posterior pituitary occurs in response to signals from the nervous system. Sensory signals are transmitted along afferent nerve fibers to the hypothalamus. From the hypothalamus neural information is transmitted to the posterior pituitary resulting in the release of the specific hormone. The secretion of hormones from the posterior pituitary that are involved in mammalian reproduction will be discussed later.

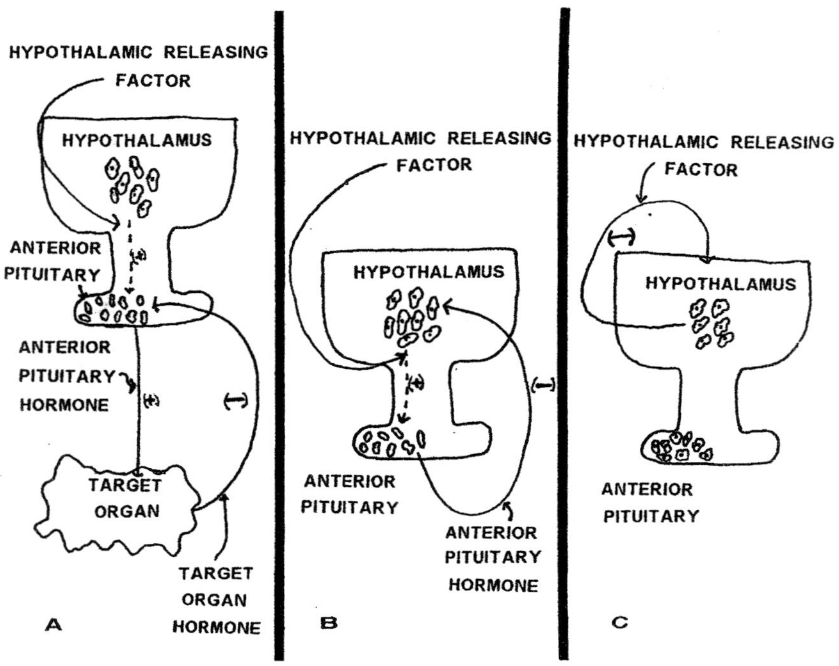

Figure 2.9. General mechanism illustrating negative feedback systems in the hypothalamic-pituitary-target organ axis. (A) Long loop negative feedback system. (B) Short loop negative feedback system. (C) Ultrashort loop negative feedback system.

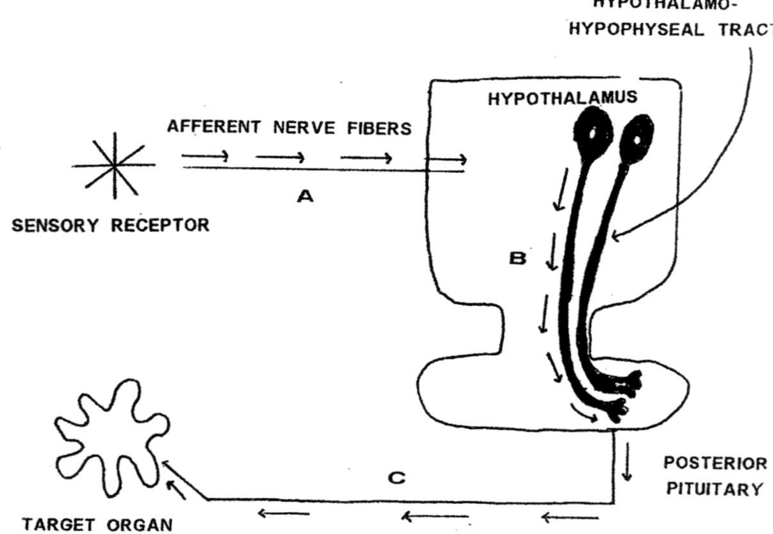

Figure 2.10. General mechanism illustrating a neuroendocrine reflex. (A) Sensory signals are detected by a receptor. Bioelectrical information is transmitted along afferent nerve fibers to the hypothalamus. (B) Neurons in the hypothalamus synthesize a hormone that is transported via the hypothalamo-hypophyseal tract to the posterior pituitary. (C) Posterior pituitary secretes the hypothalamic hormone into the blood. The blood transports the hormone to the specific target organ. Hypothalamic hormone changes the physiology of the target organ.

FUNCTIONAL ANATOMY OF THE MAMMALIAN MALE REPRODUCTIVE SYSTEM

FUNCTIONAL ANATOMY OF THE TESTES

<u>Testicular Location and Scrotal Anatomy</u>

The **testes** of most mammalian species are located either in the abdominal cavity or in an outpocketing of the abdominal wall called the **scrotum** (Figure 3.1). The elephant, hyrax and dolphin are examples of animals with abdominal testes. Their testes are located in the posterior end of the abdominal cavity. Cattle, sheep, swine, rabbits and humans are examples of animals with scrotal testes. The scrotum is pendulous in cattle, sheep and humans whereas the testes are located in a nonpendulous scrotum in swine and rabbits.

The scrotum is divided into two lateral compartments by the medial ridge (Figure 3.2). The septum is composed of fascia and the **dartos muscle**. Contraction of the dartos muscle during periods of external cool temperatures causes wrinkling of the scrotal skin. The **cremaster muscle** is a band of skeletal muscle that attaches to the testes and regulates the position of the testes within the scrotum. Contraction of this muscle elevates the testes bringing them closer to the abdominal cavity.

<u>Testicular Anatomy</u>

Several layers of nonfunctional tissue (Figure 3.3) cover the testicle. The **tunica vaginalis**, a continuation of the peritoneum, is a serous membrane that surrounds the testicle. The **tunica albuginea** (testicular capsule), a layer of dense fibrous connective tissue, is located beneath the tunica vaginalis. The tunica albuginea in some species of mammals contains smooth muscle cells and may be important in maintaining appropriate pressure within the testicle.

Connective tissue from the tunica albuginea invades the internal parenchyma of the testicle dividing it into 200 to 300 lobules. The functional parenchyma within each lobule can be divided into the **seminiferous tubules** and the **testicular interstitium**. The diameter of the seminiferous tubules in adult mammalian males is usually between 200 to 300 micrometers and comprises from 60% to 90% of the testicular mass. The number of seminiferous tubules is species dependent. Most seminiferous tubules are open at both ends and empty into the rete testis. The cell types found within the seminiferous tubules include the **Sertoli cells** and the developing germ cells. The interstitium is the area between the seminiferous tubules. Within the testicular interstitium are nerves, blood and lymphatic vessels, and **Leydig cells.**

The testicle has two duct systems called the **rete testis** and the **efferent ducts**. These duct systems convey spermatozoa and fluids from the seminiferous tubules to the epididymis. In rams, bulls and boars the rete testis is centrally located within the testicle. The rete testis of humans and rodents is located at the side of the testicle near the epididymis. The ends of the seminiferous tubules enter into the rete testis. The epithelium of the rete testis is composed of cuboidal cells in which some cells are ciliated. These ciliated cells are believed to aid in the transport of immotile spermatozoa into the efferent ducts. The efferent ducts (3 to 8 ducts) lead from the rete testis into the epididymis where they converge to form the epididymal duct. The epithelium of the efferent ducts contains some ciliated columnar cells as well as short nonciliated cells. Ciliary action propels immotile spermatozoa into the epididymis.

FUNCTIONAL ANATOMY OF THE MAMMALIAN MALE REPRODUCTIVE SYSTEM

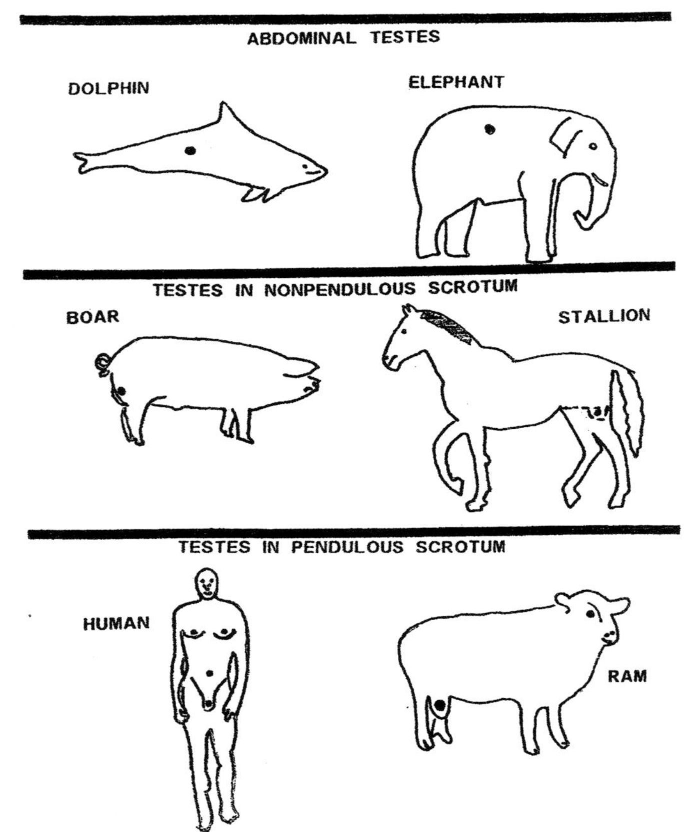

Figure 3.1. Testes location in different mammalian species.

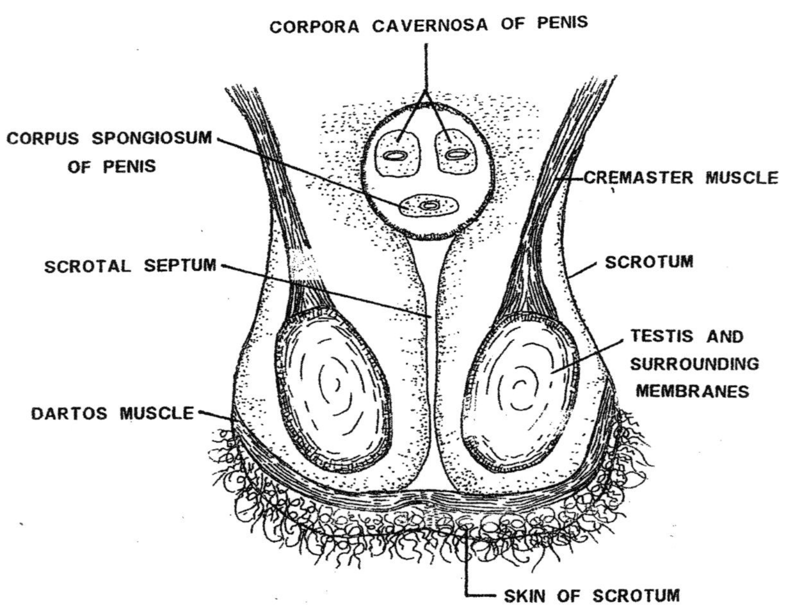

Figure 3.2. Anatomy of a pendulous scrotum.

FUNCTIONAL ANATOMY OF THE MAMMALIAN MALE REPRODUCTIVE SYSTEM

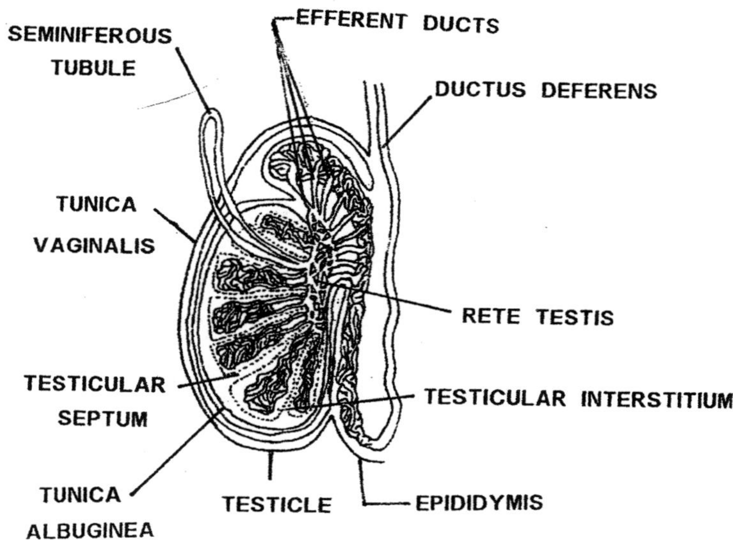

Figure 3.3. General anatomy of a mammalian testicle.

The Diploid Cells of the Seminiferous Tubule

The somatic cells of the seminiferous tubule are the **peritubular myoid cells** and the **Sertoli cells**. The peritubular myoid cells are smooth muscle cells that comprise the boundary of the seminiferous tubule external to the basement membrane. A peritubular myoid cell has a central elliptical shaped nucleus, sparse cytoplasmic organelles and abundant microfilaments. Peritubular myoid cells contribute to the blood-testis barrier by preventing the passage of high molecular weight substances from the testicular interstitium into the seminiferous tubule. Peritubular myoid cells are responsible for peristaltic type movements within the seminiferous tubule. Peritubular myoid cells have been shown to produce **leukemia inhibiting factor (LIF)** and a protein called **PMod S**. PModS has been shown to have a stimulatory effect on Sertoli cells. Leukemia inhibitory factor may have a stimulatory effect on Sertoli cells and an influence on spermatogenesis. The basement membrane of the seminiferous tubule contains laminin, type IV collagen, heparin sulphate and proteoglycan. The basement membrane provides structural support for the seminiferous tubule and is involved in morphological changes of cells located in the basal compartment of the seminiferous tubule.

Sertoli cells are found within the seminiferous tubule. Sertoli cells extend from the basement membrane of the seminiferous tubule to the lumen. Sertoli cells surround the developing germ cells (Figure 3.4). Sertoli cells are irregular in shape and each cell has an indented nucleus. Sertoli cells have numerous mitochondria and are richly endowed with smooth endoplasmic reticulum and rough endoplasmic reticulum. The smooth endoplasmic reticulum of the Sertoli cell contains a number of lipid droplets that varies among species and with the state of the seminiferous epithelial cycle. Sertoli cells are rich in Golgi

FUNCTIONAL ANATOMY OF THE MAMMALIAN MALE REPRODUCTIVE SYSTEM

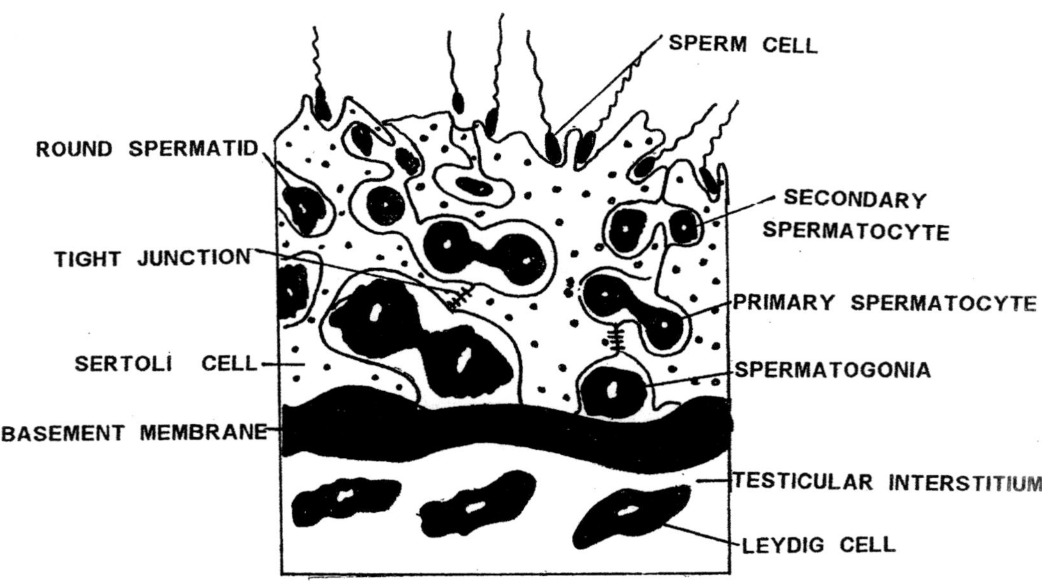

Figure 3.4. Anatomy of the seminiferous tubule.

bodies and have large vacuoles and exocytotic vesicles.

<u>Junctional Complexes of Sertoli cells</u>

Sertoli cells form **tight junctions** among themselves. Tight junctions between Sertoli cells prevent the passage of blood elements from the basal compartment of the seminiferous tubule to the adlumenal compartment of the seminiferous tubule. This **blood-testis barrier** develops in most species of mammals just prior to the onset of spermatogenesis and prevents immunological sensitivity to haploid germ cells. The barrier also prevents compounds with toxicological properties from coming into contact with developing germ cells.

Sertoli cells form junctional complexes with germinal cells. **Desmosomes** form between Sertoli cells and B spermatogonia and also between Sertoli cells and pachytene spermatocytes. This type of junctional complex appears to be important in the translocation of germ cells across the blood-testis barrier. **Ectoplasmic specializations** appear between Sertoli cells and pachytene spermatocytes but are less obvious during the second half of meiosis. Ectoplasmic specializations reappear during the time of spermatid formation. The function of this type of junction is to maintain the tight relationship between the Sertoli cells and the developing germ cells. **Tubular-bulbar complexes** (Figure 3.5) are bristle-coated pits in which projections of spermatid cytoplasm extend to form bulbous dilations at the surface of Sertoli cells. When spermatogenesis is complete, the Sertoli cells withdraw from the heads of the spermatozoa causing breakage of the tubular-bulbar complexes. This releases the spermatozoa into the lumen of the

FUNCTIONAL ANATOMY OF THE MAMMALIAN MALE REPRODUCTIVE SYSTEM

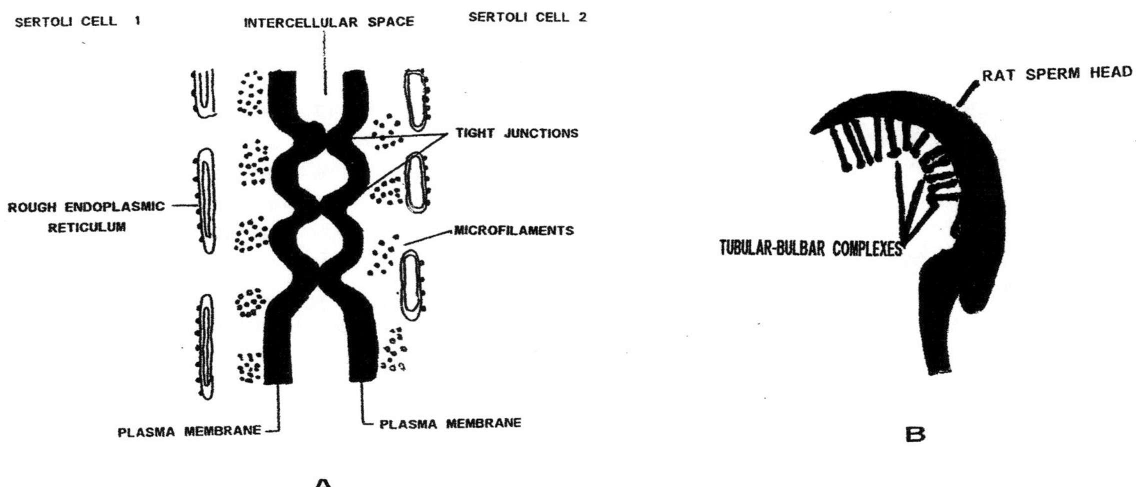

Figure 3.5. Sertoli cell junctions found within the seminiferous tubule. (A) Tight junction between two Sertoli cells. (B) A tubular-bulbar complexes between a spermatozoon and a Sertoli cell.

seminiferous tubule.

<u>Fluid Composition in the Seminiferous Tubule and Rete Testis</u>

Sertoli cells produce fluid that enters the rete testis via the seminiferous tubule. Seminiferous tubule fluid of most mammalian species has a pH of approximately 7.3. Seminiferous tubule fluid is higher in potassium, bicarbonate, inositol and glutamic acid but lower in sodium, chlorine, protein and glucose when compared to blood. Some of the protein found in the seminiferous tubular fluid includes mitotic stimulators, acrosin inhibitors, inhibin and androgen binding protein. The composition of the seminiferous tubular fluid changes as it enters the rete testis. For example, rete testis fluid is higher in sodium and chlorine, but lower in potassium and bicarbonate when compared to seminiferous tubule fluid. Tight junctions exist between the epithelial cells of the rete testis allowing the rete testis to maintain a unique fluid composition.

<u>Spermatogenesis</u>

Spermatogenesis is the formation of spermatozoa. The seminiferous tubule is the site of spermatogenesis (Figure 3.4). Light microscopic examination of the seminiferous tubule reveals four types of germinal cells: **gonocytes, spermatogonia, spermatocytes** and **spermatids**.

Gonocytes are primitive diploid reproductive cells that are surrounded by Sertoli cells in the middle of the seminiferous tubule during prepubertal development. Gonocytes remain mitotically inactive until shortly before puberty. At puberty, gonocytes migrate between the Sertoli cells to reach the basement membrane of the seminiferous tubule. Once mitosis starts, gonocytes are referred to as spermatogonia.

During the process of **spermatocytogenesis** (mitotic division of spermatogonia), gonocytes differentiate into A_0 **spermatogonia**. From the mitotic division of A_0 spermatogonia A_1, A_2, and A_3 spermatogonia arise. The A_2 **spermatogonia** mitotically divide forming dormant A_1 **spermatogonia** (for later cellular divisions) and active A_3 **spermatogonia**. Mitotic division of A_3 spermatogonia give rise to **intermediate spermatogonia**, which then divide forming B_1 **spermatogonia**. Mitotic division of B_1 spermatogonia gives rise to B_2 **spermatogonia** that divide forming **primary spermatocytes**. In summary, one active A_3 spermatogonia gives rise to 16 primary spermatocytes. In bulls and rams, spermatocytogenesis is completed within 15 days to 17 days. Each primary spermatocyte undergoes meiotic division. During the first half of meiosis, each primary spermatocyte forms two haploid **secondary spermatocytes**. Within a few hours each secondary spermatocyte enters the second half of meiosis yielding two **round spermatids**. In the ram, meiosis lasts 15 days to 17 days.

In **spermiogenesis**, the round spermatid is transformed into a spermatozoon (Figure 3.6). Spermiogenesis requires 12 days to 14 days in the rat and approximately 23 days in the human male. During spermatogenesis, the nuclear chromatin condenses forming the head of the sperm cell as the rest of the cell elongates forming the tail. The **acrosome** (a sac containing enzymes) forms on the head of future spermatozoon. One centriole forms the axoneme of the tail, whereas the other centriole forms the neck that attaches the nucleus to the tail. The cytoplasm is cast off during the formation of the spermatozoon's tail forming a cytoplasmic droplet around the neck of the cell. Mitochondria gradually wrap themselves around the mid-piece of the spermatozoon forming the **mitochondrial sheath**. Finally, the tubular-bulbar complex breaks releasing the spermatozoon into the seminiferous tubule lumen.

Acrosomal formation can be divided into 3 phases, the **Golgi phase**, the **cap phase**, and the **acrosomal phase**. In the Golgi pahase, the Golgi apparatus gives rise to several cisternae containing granules. These proacrosomic granules fuse to form a single spherical granule that attaches to the nuclear envelope of the future spermatozoon. During the cap phase, the spherical acrosomic granule enlarges by the addition of glycoproteins. The acrosome granule starts to flatten over the head of the future spermatozoon. In the acrosomal phase, the flat acrosomic granule attaches to the head of the future spermatozoon.

<u>Spermatogenic Cycle and the Spermatogenic Wave</u>

 If a person were to microscopically view a portion of the seminiferous tubule, they would notice a certain association of germinal cells. If this area of the seminiferous tubule was observed using time lapse photography, a series of different germinal cell associations would be seen until a cycle was complete. This phenomenon of the seminiferous tubule is referred to as the **spermatogenic cycle** (Figure 3.7). The length of seminiferous tubule showing particular cell associations varies among species. The duration of the spermatogenic cycle is species dependent. In humans, bulls, stallions, rams and boars the spermatogenic cycle is 16, 13.5, 12.2, 10.3 and 8.6 days in length, respectively.

The **spermatogenic wave** is the sequence of germinal cell associations found along the length of the seminiferous tubule at a given moment in time (Figure 3.8). To see the spermatogenic wave, a person must travel along the length of the seminiferous tubule beginning at the rete testis. Successive lengths of the tubule would be at an earlier stage of the spermatogenic cycle forming a wave of spatial arrangements of in a temporal cycle. Because the seminiferous tubule is open at both ends, a reversal in the spatial pattern of

FUNCTIONAL ANATOMY OF THE MAMMALIAN MALE REPRODUCTIVE SYSTEM

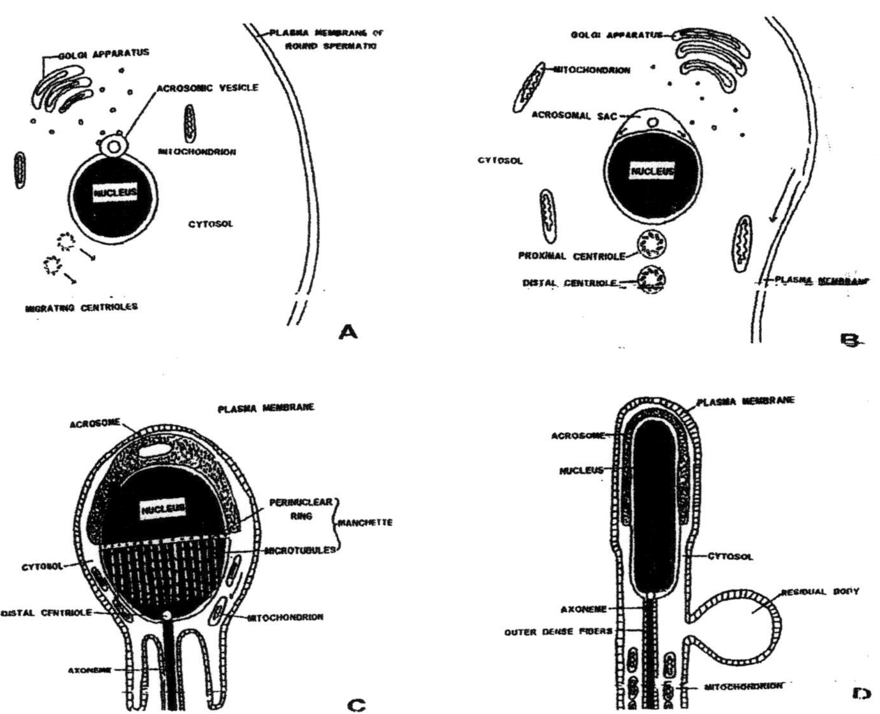

Figure 3.6. Spermiogenesis and acrosome formation in the mammalian testis. (A) Golgi phase of acrosome formation. (B) Cap phase of acrosome formation. (C) Acrosome phase of acrosome formation. (D) Developed spermatozoon.

LUMEN OF THE SEMINIFEROUS TUBULE

		L	L	L	L	L	
R	EL	D	C2	R	R	R	L
P	P	Z	Z	Z	P	P	R
B, LE	Z	A3	IN	IN	B1	B1	P
A1	A2	A1	A1	A1	A1	A1	B1 B2
A0	A0	A0	A0	A0	A0	A0	A0
1	2	3	4	5	6	7	8

STAGES OF THE SPERMATOGENIC CYCLE

Figure 3.7. The spermatogenic cycle. A0 and A1 = reserve and renewing stem cells, respectively. A2 and A3 = type A spermatogonia. IN = intermediate spermatogonia. B1 and B2 = type B spermatogonia. LE, Z, P, D = leptotene, zygotene, pachytene and diplotene primary spermatocytes, respectively. C2 = secondary spermatocytes. R, EL, and L = round spermatid, elongating spermatid and elongated spermatid, respectively.

FUNCTIONAL ANATOMY OF THE MAMMALIAN MALE REPRODUCTIVE SYSTEM

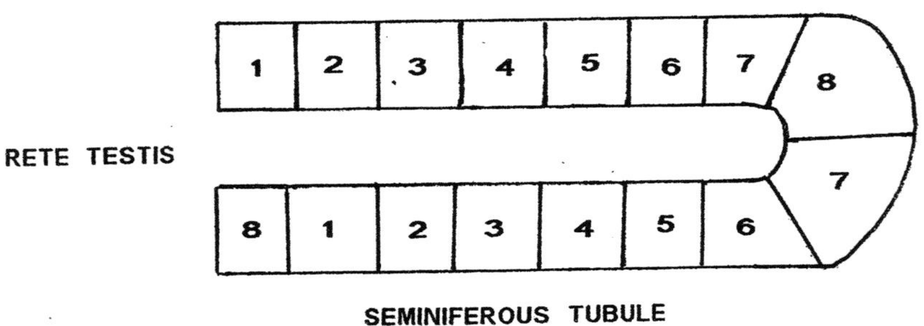

Figure 3.8. Spermatogenic wave in a seminiferous tubule. A sequence of germinal cell associations (1 – 8) found along the length of the seminiferous tubule at a given moment of time.

reversal in the spatial pattern of germinal cell associations occurs along the length of the seminiferous tubule. How the spermatogenic wave originates is not known, but may be a consequence of the way spermatogenesis begins at puberty. Once the spermatogenic wave is established, it persists because of the synchronized development of the different germ cells.

<u>Factors that Influence Spermatogenesis</u>

Mammals with testes in a pendulous scrotum have testicular temperatures 3 C to 4 C cooler than core body temperature. If the testicular temperature is equal or greater than core body temperature (e.g. cryptorchism), spermatogenesis does not occur. Microscopic examination of the seminiferous tubule reveals only the presence of spermatogonia and Sertoli cells with few spermatocytes and spermatids. It is unknown why the testes cannot function at core body temperature. Hypoxia induced by an increase in cellular metabolism without a corresponding change in blood flow may be responsible.

Dietary deficiencies can also have an effect on spermatogenesis. If food intake is severely restricted, spermatogenic function is drastically decreased. This is partially due to a decrease in the synthesis and secretion of gonadotropic hormones. Deficiencies in **vitamin A**, **vitamin E**, essential fatty acids, certain amino acids and **zinc** also drastically decrease spermatogenesis. This appears to be due to direct effects on the testes, since pituitary functions are not altered.

FUNCTIONAL ANATOMY OF THE MAMMALIAN MALE REPRODUCTIVE SYSTEM

Certain chemicals can alter seminiferous tubule function. Excessive intake of **cadmium salt** causes testicular degeneration. Cytotoxic drug **busulphan**, **methotrexate**, **vinblastine** and **vincristine** are toxic to spermatozoa. **Hydroxyurea** can alter DNA synthesis of germinal cells. **Ethylene dibromide** and **dibromochloropropane** can reduce the number of sperm cells in the ejaculate. Men often become infertile after contracting **mumps**. The cause of this infertility is unknown, but may be the result of inflammation, increased testicular temperature or an immunological reaction causing the breakdown of the blood-testis barrier.

<u>Anatomy of the Spermatozoon</u>

The sperm cell can be divided into two parts called the head and the tail (Figure 3.9). The shape of the head can vary between species. For example, spermatozoa of rats and mice have hooked heads, whereas those of humans and domestic animal species are rounded (Figure 3.10). The sperm cell head contains a condensed mass of chromatin (DNA and protein) (Figure 3.11). Disulfide bonds stabilize the chromatin until fertilization. An enclosed membrane called the **acrosome** covers the anterior portion of the head. The shape of the acrosome is species dependent. The **post acrosomal region** is inferior to the acrosome. The post acrosomal region is the portion of the head that binds to the egg during fertilization. A thick dense layer called the **post acrosomal sheath** underlies the plasma membrane in the post acrosomal region. The space between the sheath and the nucleus extends back to the posterior ring where the plasma membrane is fused with the underlying **nuclear envelope**. This portion of the nuclear membrane has no pores. The part of the sperm head near the mid-piece forms the **implantation fossa**. A thick layer of dense material that constitutes the **basal plate** covers this region of nuclear membrane.

The **sperm tail** is divided into three regions called the **mid-piece, principal piece**, and the **end piece** (Figure 3.9). At the center of the tail is the **axoneme**, which consists of fibers of microtubules (Figure 3. 12). The axoneme contains a central fiber (one pair of microtubules) surrounded by nine spaced inner fibers. Each inner fiber is a microtubule doublet with one microtubule called subfiber A and the other subfiber B. Subfiber A has two dynein arms projected in the direction of the next adjacent inner fiber. These dynein arms have strong ATPase activity and are responsible for sperm motility. There are also slender nexin links joining the adjacent inner fibers. Radial spokes join each inner fiber to the helical sheath around the central fiber. This axonemal complex is continuous throughout the tail.

The mid-piece extends from the head of the sperm cell to the end of the **mitochondrial helix**. The portion of the mid-piece nearest the head is called the **connecting piece** (Figure 3.11). At the proximal end of the connecting piece is the articulation region called the **capitulum**. The capitulum is attached to the nucleus by a series of fine filaments. Extending back from the capitulum are **nine dense outer fibers** that overlap the axoneme (Figure 3.12). Outside the dense outer fibers, mitochondria wrap around the mid-piece. There is considerable species variation in the number of gyres in the mitochondrial helix. A **fibrous sheath** covers the dense outer fibers of the principal piece (Figure 3.12). The fibrous sheath consists of a number of circumferentially oriented running ribs between the longitudinal columns that extend along the entire length of the principal piece. As the principal piece tapers the outer fibers become smaller and the fibrous sheath becomes thinner.

The **end piece** is the last several microns of the sperm tail. This portion of the sperm tail is the axoneme covered by a plasma membrane (Figure 3.13).

FUNCTIONAL ANATOMY OF THE MAMMALIAN MALE REPRODUCTIVE SYSTEM

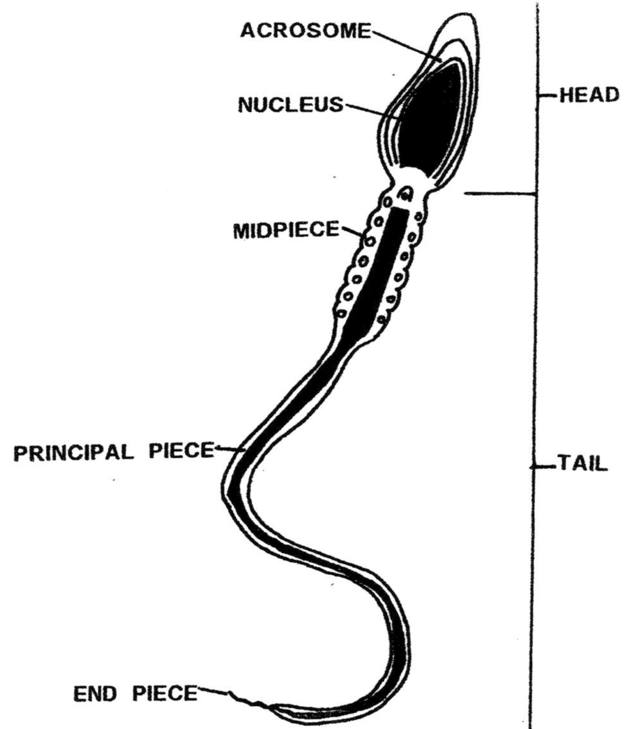

Figure 3.9. The anatomy of a spermatozoon.

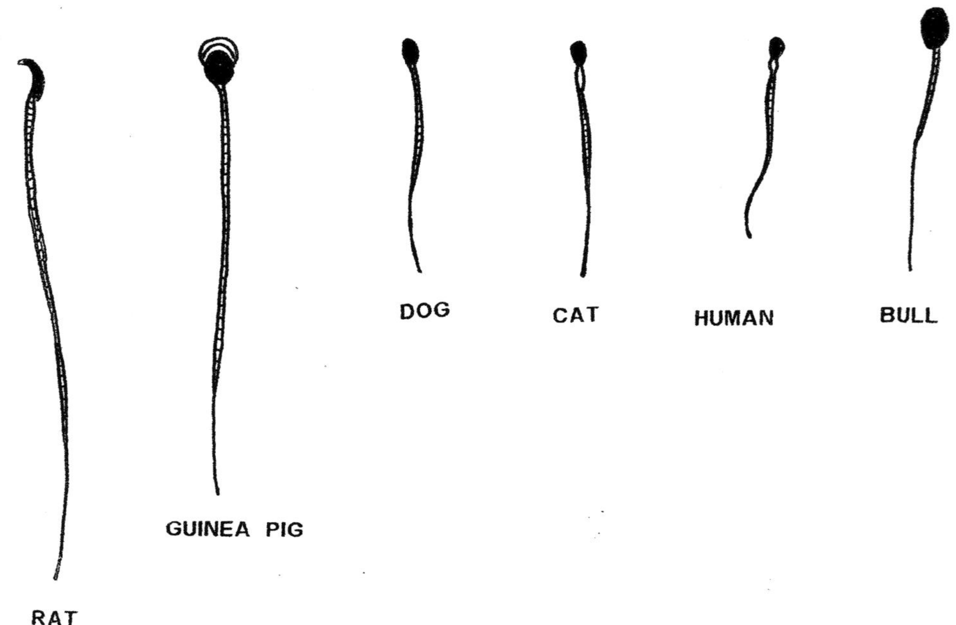

Figure 3.10. Spermatozoon anatomy of several mammalian species.

FUNCTIONAL ANATOMY OF THE MAMMALIAN MALE REPRODUCTIVE SYSTEM

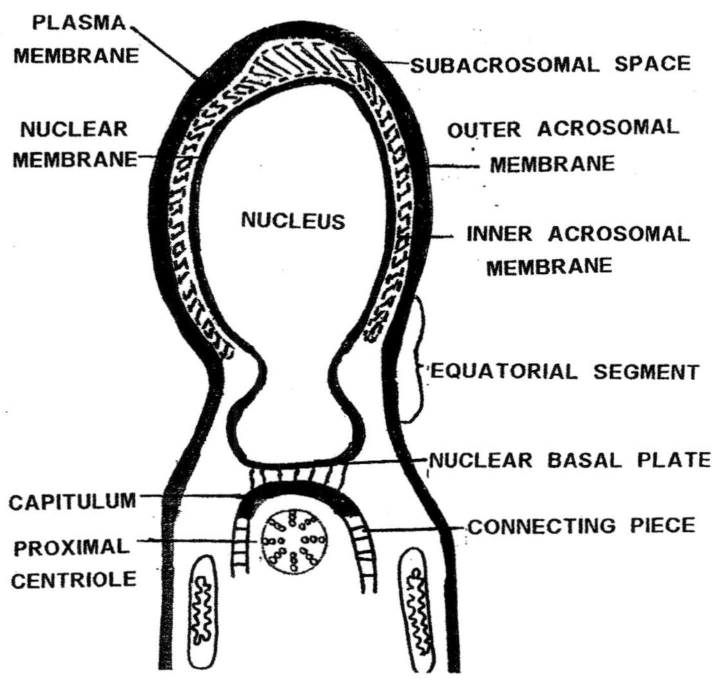

Figure 3.11. Electron microscopic anatomy of a sperm cell head.

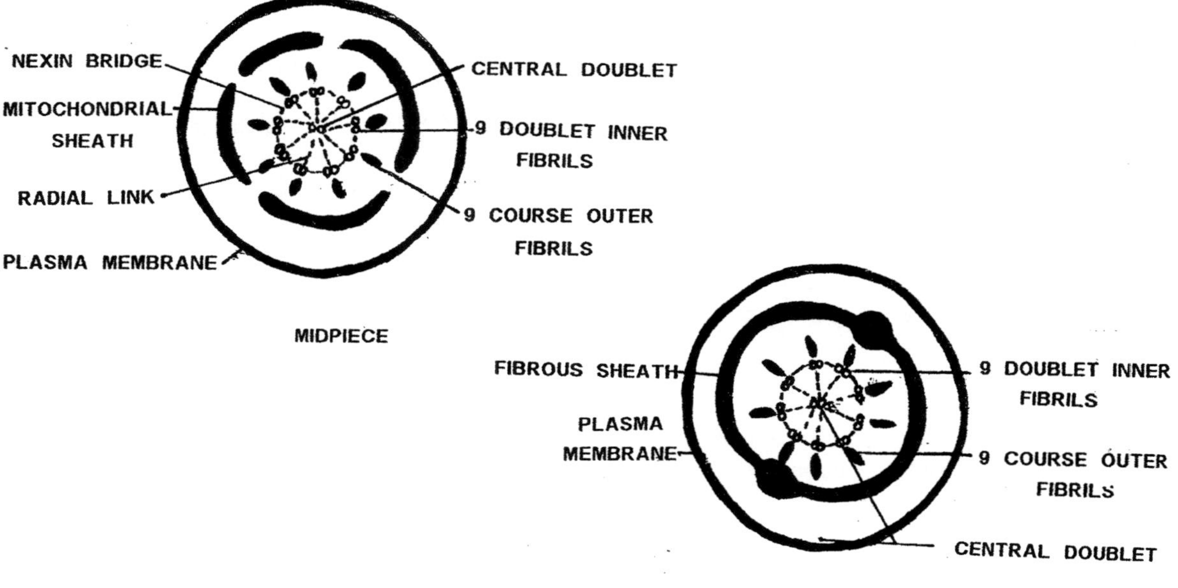

Figure 3.12. Electron microscopic anatomy of the mid-piece and principal piece of a sperm tail.

FUNCTIONAL ANATOMY OF THE MAMMALIAN MALE REPRODUCTIVE SYSTEM

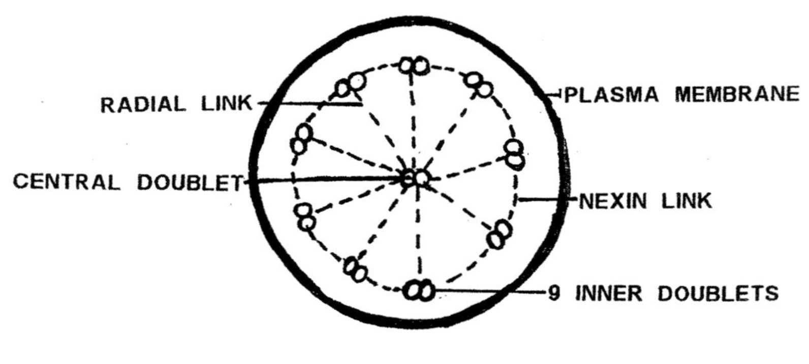

Figure 3.13. Electron microscopic anatomy of the end piece of a sperm tail.

Acrosomal Enzymes

The acrosome contains numerous enzymes that are involved in the process of fertilization. **Glycohydrolases** are a group of intraacrosomal enzymes and their functions in fertilization are unknown. It is possible that the glycohydrolases are involved in the penetration of the corona radiata of the secondary oocyte. The **proteinases/proteases** are present in the acrosome and on the surface of capacitated sperm. These enzymes are involved in the penetration of the zona pellucida surrounding the secondary oocyte and will be discussed in more detail later. **Phospholipase A₂** and **phospholipase** are involved in the signal transduction pathway during acrosomal exocytosis. **Acid phosphatase** and **alkaline phosphatase** are involved in the acrosome reaction. **Aryl sulphatases** have been found in the acrosomal extracts of many domestic animals. The function of this group of enzymes in the fertilization process is unknown.

Sperm Cell Metabolism

Spermatozoa have the ability to utilize a wide range of substrates for energy. Simple sugars such as glucose, fructose and mannose can be reduced anaerobically to lactic acid or oxidized to carbon dioxide and water depending on the circumstances. Organic substances such as lactate, pyruvate and fatty acids can aerobically be used for energy. Spermatozoa are unusual in that the rate of glycolysis is not reduced under aerobic conditions.

<u>Sperm Motility</u>

The immediate source of energy for sperm motility is ATP, ADP and AMP. High cellular concentrations of carnitine can serve as a metabolic reserve for ATP.

Sperm tail motility involves the sliding of adjacent inner fibers within the axoneme. **Dynein** is responsible for converting energy into mechanical movement (Figure 3.14). In order to slide two adjacent inner fibers against each other, ATP binds to a flattened dynein arm of subfiber A. As ATP is broken down, the dynein arm begins to extend eventually coming in contact with subfiber B of the adjacent fiber. The dynein arm then adopts a rigid position as the products of ATP hydrolysis are released. The attachment of ATP to the dynein arm releases the contact with subfiber B and allows the arm to resume a flattened position against subfiber A. In the process, the fibers slide in relation to each other. The movement of the inner fibers of one half of the axoneme is in synchrony, but they are out of phase with the inner fibers in the second half of the axoneme. This lack of synchrony allows the sliding motion to be converted into a bending motion (Figure 3.15).

<u>Endocrinology of the Seminiferous Tubule</u>

Follicle stimulating hormone is delivered to the testicular interstitium by small arterioles. Follicle stimulating hormone diffuses from the testicular interstitium, across the basement membrane of the seminiferous tubule, to receptors located on the plasma membrane of Sertoli cells. In response to FSH stimulation, Sertoli cells produce a wide range of proteins (in excess of 100!) that include **androgen binding protein** (ABP) and **inhibin**. Follicle stimulating hormone is also required for the production of intracellular androgen receptor proteins in the Sertoli cells as well as the production of seminiferous tubule fluid. Androgen binding protein is secreted from Sertoli cells into the seminiferous tubule. Androgen binding protein binds to androgens that have diffused from the testicular interstitium into the seminiferous tubule. Androgen binding protein transfers androgens to the promeiotic germinal cells that have androgen receptors. Sertoli cells also secrete ABP into the seminiferous tubule fluid. This may serve to increase the amount of androgens being transported to the epididymis.

Inhibin is involved in the regulation of plasma FSH using a negative feedback mechanism (Figure 3.16). Elevation of plasma inhibin concentration causes the anterior pituitary to be less sensitive to GnRH stimulation and to synthesize and secrete less FSH. Inhibin also has a weak negative feedback effect on the hypothalamus causing a decrease in the synthesis and secretion of GnRH. Inhibin appears to have autocrine and paracrine functions within the testis. Inhibin receptors are found on the Leydig cells. Elevation of testicular inhibin is directly related to the completion of spermiogenesis in humans.

Follicle stimulating hormone is involved in the stimulation and maintenance of spermatogenesis. Because the effects of FSH are mediated via the Sertoli cell, the mechanism by which FSH influences spermatogenesis is unclear and appears to differ with age, season and species.

Prolactin controls the yield of spermatozoa during spermatogenesis. The mechanism by which this occurs is unclear. Receptors for prolactin are found on Leydig cells. This indicates that prolactin influences spermatogenesis indirectly.

FUNCTIONAL ANATOMY OF THE MAMMALIAN MALE REPRODUCTIVE SYSTEM

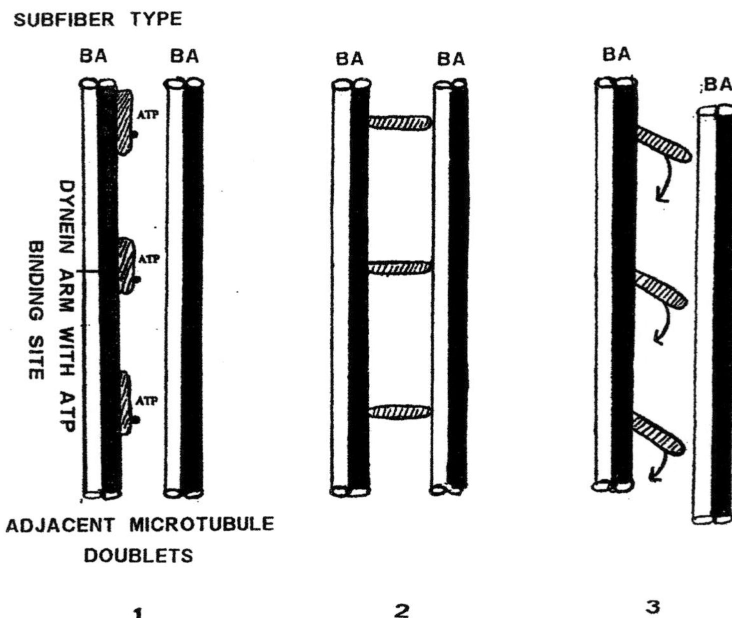

Figure 3.14. Mechanism of sperm cell motility. 1. ATP binds to the dynein arm of subfiber A. 2. As ATP is broken down, the dynein arm extends and contacts subfiber B of the adjacent fiber. 3. The attachment of ATP to the dynein arm releases the arm's contact with subfiber B and the dynein arm resumes a flattened position against subfiber A.

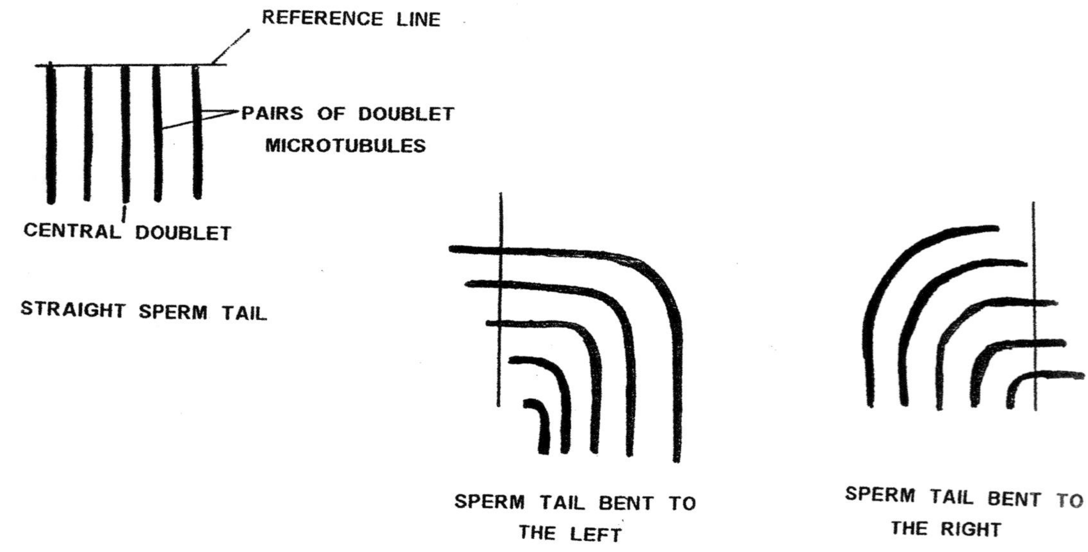

Figure 3.15. Mechanism of sperm tail bending. Movement of the inner fibers of one half of the axoneme is in synchrony, but are out of phase with the inner fibers of the second half of the axoneme. This lack of synchrony allows the sliding motion of the fibers to be converted into a bending motion.

FUNCTIONAL ANATOMY OF THE MAMMALIAN MALE REPRODUCTIVE SYSTEM

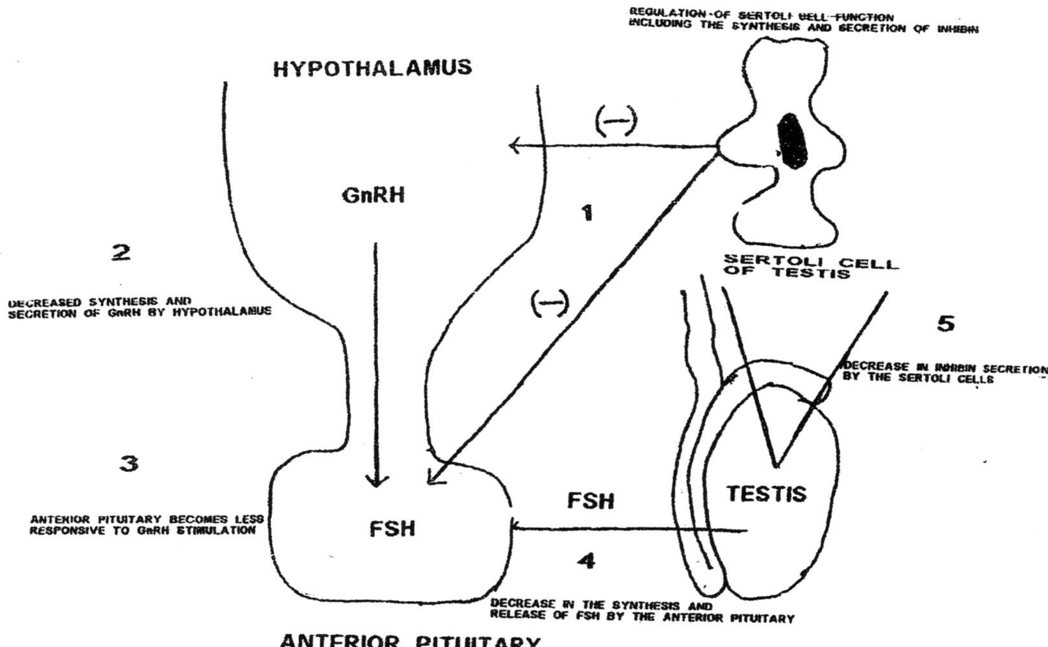

Figure 3.16. Negative feedback regulation of plasma FSH concentration by testicular inhibin 1. Hypothalamus produces and secretes GnRH into the hypothalamic-hypophyseal portal system. 2. GnRH stimulates the anterior pituitary to produce and secrete FSH into the blood. 3. FSH stimulates the Sertoli cells of the testes to secrete inhibin into the blood. 4. Elevated plasma levels of inhibin causes the hypothalamus to produce and secrete less GnRH. 5. Elevated plasma levels of inhibin cause the anterior pituitary to be less responsive to GnRH stimulation therefore decreasing the synthesis and release of FSH. 6. A decrease in plasma levels of FSH results in a decrease in the synthesis and release of inhibin by the testes.

Testosterone and androstenedione from the testicular interstitium diffuses across the basement membrane of the seminiferous tubule and into the Sertoli cell (Figure 3.17). Most androgens are converted to dihydrotestosterone by 5 alpha-reductase. Dihydrotestosterone and a small amount of testosterone that is secreted into the seminiferous tubule fluid bind to androgen binding protein. Androgen receptor proteins make the Sertoli cell a target for androgen stimulation. Androstenedione and testosterone in the Sertoli cell can be converted to estrogens (androstenedione to estrone and testosterone to estradiol -17B). In some mammalian species (e.g. stallion), these estrogens and secreted in large amounts into the blood.

<u>Testicular Interstitial Anatomy and Leydig Cell Development</u>

The testicular interstitium is the area of the testes between the seminiferous tubules. Within this area are nerves, blood and lymphatic vessels, and Leydig cells.

The **Leydig cell** is the principal cell type found in the testicular interstitium. Leydig cells are relatively large and polyhedral in shape with round eccentrically placed nuclei. The cytoplasm of Leydig cells contains considerable smooth endoplasmic reticulum and moderate numbers of mitochondria. The Golgi complex is well developed. Leydig cells of some mammalian species (e.g. boar) show large number of

FUNCTIONAL ANATOMY OF THE MAMMALIAN MALE REPRODUCTIVE SYSTEM

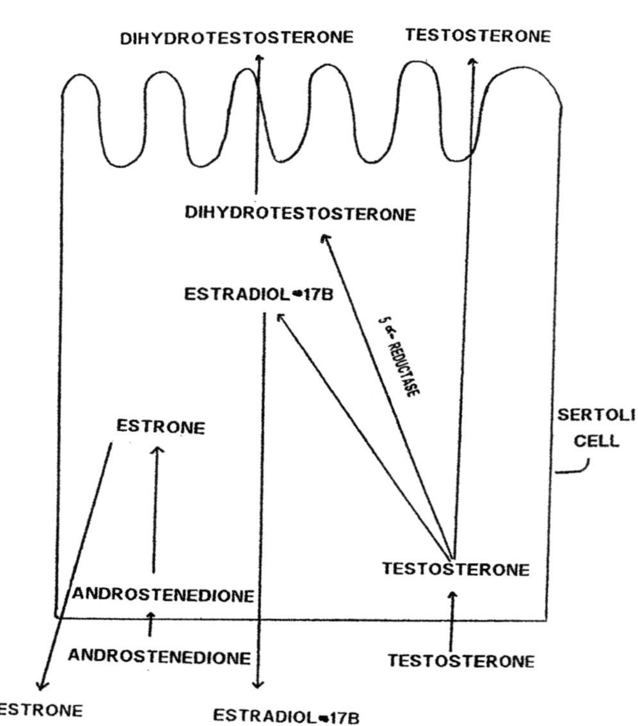

Figure 3.17. Steroid production in the Sertoli cell.

lipid droplets in the smooth endoplasmic reticulum. These droplets contain esterized cholesterol. The relative mass of Leydig cells varies between species. In bulls and rams, Leydig cells comprise approximately 1% of the testicular mass, whereas in the stallion and boar Leydig cells comprise approximately 28% and 37% of the testicular mass, respectively.

The number of Leydig cells change during the animal development. In boars, Leydig cells develop from undifferentiated mesenchymal cells during early gestational development and produce high levels of androgens. Elevated androgen levels are responsible for establishing the male sex ducts. At midgestation the number of Leydig cells decrease in the boar. Leydig cells in the boar testis increases again several weeks prior to birth only to regress approximately three weeks post parturition. Androgen levels in the plasma of the boar are elevated during this time period. At puberty, Leydig cell development once again occurs in the boar with a corresponding increase in plasma androgen levels.

Leydig cell numbers also show seasonal cycles in some species of mammals (e.g. deer). In these species, Leydig cell numbers increase and plasma androgen levels become elevate only during the breeding season.

<u>Endocrinology of the Testicular Interstitium</u>

The Leydig cell is the major site of steroid synthesis in the testes. Esterized cholesterol in the smooth endoplasmic reticulum is moved to mitochondria by cytoskeletal microfilaments. **Cholesterol** is then

FUNCTIONAL ANATOMY OF THE MAMMALIAN MALE REPRODUCTIVE SYSTEM

transported from the outer membrane of the mitochondrion to the inner membrane. The cholesterol side chain is cleaved by enzyme $P450_{scc}$ converting cholesterol to **pregnenolone**. Pregnenolone is then transported to the smooth endoplasmic reticulum. In the presence of 17 alpha-hydrolase, 17-20 lyase, and 3 beta-hydroxysteroid dehydrogenase, pregnenolone is converted to **testosterone**. (Figure 3.18).

Androgen synthesis by Leydig cells is primarily the result of LH stimulation (Figure 3.19). Gonadotropic releasing hormone is secreted by the hypothalamus into the hypothalamo-hypophyseal portal system and transported to the anterior pituitary. The anterior pituitary in response to GnRH stimulation secretes LH into the blood. Luteinizing hormone binds to receptors on the Leydig cells stimulating the synthesis and secretion of androgens (primarily testosterone and androstenedione) into the blood. Androgen receptors are found in the hypothalamus and the anterior pituitary. Elevation of plasma testosterone concentration decreases the synthesis and secretion of GnRH by the hypothalamus. The anterior pituitary also becomes less responsive to GnRH stimulation resulting in a decrease in the synthesis and release of LH into the blood. A decrease in plasma LH concentration results in a decrease in secretion of testosterone into the blood. Testosterone also decreases the synthesis and secretion of FSH into the blood although its effects are less when compared to LH. A more extensive suppression of FSH by the anterior pituitary comes from the combined action of testosterone and inhibin.

Leydig cells also produce several other products. **Relaxin-like factor (RLF)** is one of the most abundant products produced by Leydig cells. Deletion of the gene responsible for RLF production causes a disruption of spermatogenesis in the rat. Exactly how RLF is involved in spermatogenesis in the rat is unclear. The Leydig cell also produces oxytocin. Testicular oxytocin has been shown to act on peritubular myoid cells stimulating seminiferous tubule motility. Absorption of oxytocin into the lymphatic system and its transport to the epididymis may also be a factor in stimulating epididymal motility.

FUNCTIONAL ANATOMY OF THE EPIDIDYMIS

<u>Anatomy of the Epididymis</u>

The **epididymis** is an extremely long and convoluted duct that extends from the posterior aspect of the testicle to the lower pole where it becomes the ductus deferens. The epididymis can be divided into three regions, the **caput** (head), **corpus** (body), and **cauda** (tail). The caput and the cauda are enlarged regions, whereas the corpus is the thin region in the middle (Figure 3.20). The epididymis is surrounded by smooth muscle. From the proximal to the distal end of the epididymis, the muscle wall increases from a single circular layer to three organized layers. Slow rhythmical contractions of these muscles move the spermatozoa to the ductus deferens. At the distal end of the epididymis, sympathetic nerves that elicit intense contractions during the ejaculation process richly innervate the smooth muscle tissue.

The epithelial lining of the epididymis appears as pseudostratified. Principal cells, basal cells, narrow cells, clear cells, and halo cells comprise the epididymal epithelium. **Principal cells** have long apical microvilli. Each principal cell also contains a basal nucleus and small cytoplasmic vesicles. In the rat, principal cells comprise approximately 80% of the epithelium in the caput epididymis decreasing to 65% in the cauda epididymis. Principal cells are involved in the absorption of vast amounts testicular fluids. Tight junctions between adjacent principal cells form the **blood-epididymis barrier.** This barrier is partially responsible

FUNCTIONAL ANATOMY OF THE MAMMALIAN MALE REPRODUCTIVE SYSTEM

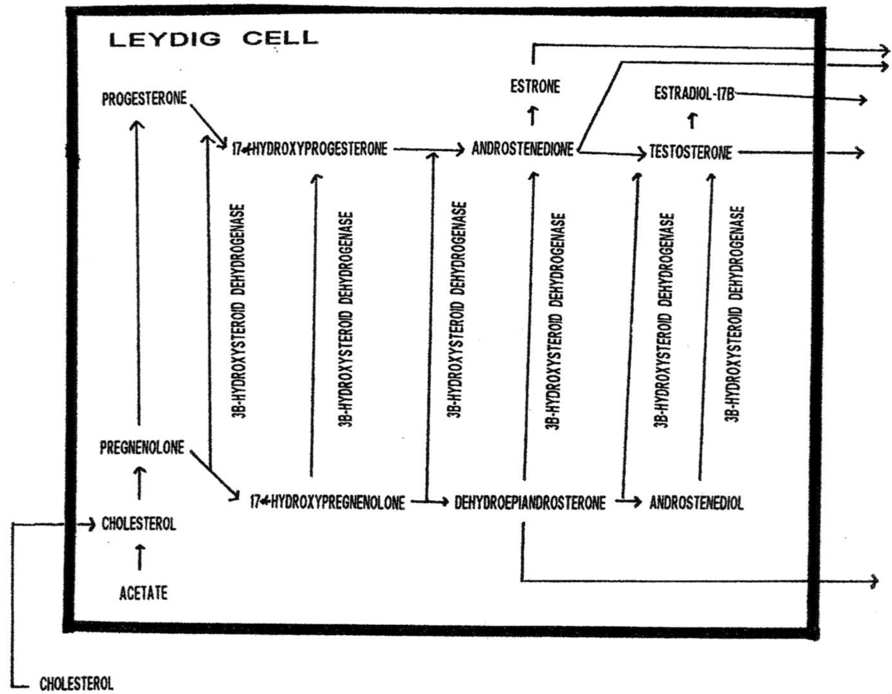

Figure 3.18. Steroid production in the Leydig cell.

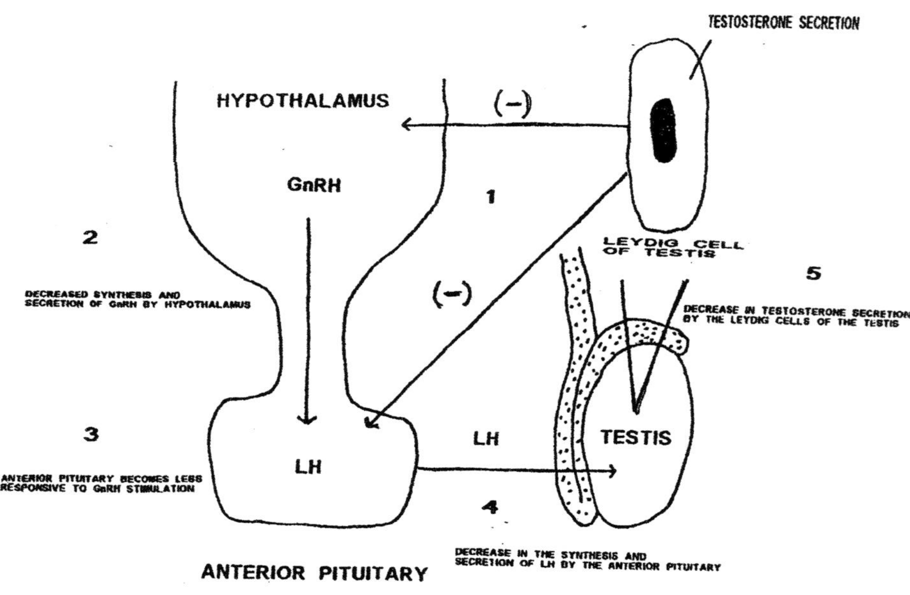

Figure 3.19. Negative feedback regulation of testicular testosterone production in the male.

FUNCTIONAL ANATOMY OF THE MAMMALIAN MALE REPRODUCTIVE SYSTEM

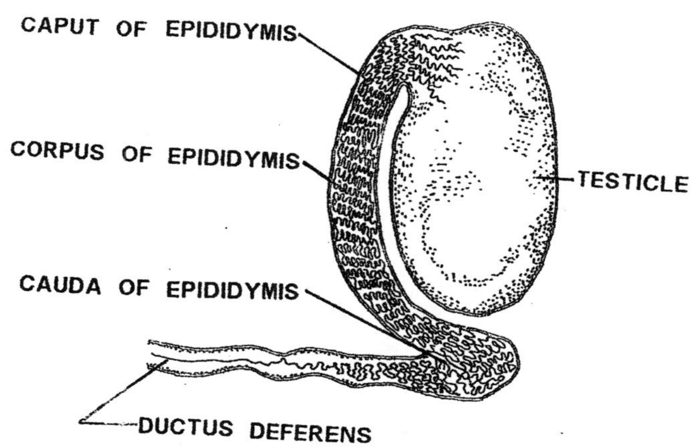

Figure 3.20. Gross anatomy of the epididymis.

for maintaining the difference in composition between epididymal luminal fluid and epididymal extraluminal fluid. **Basal cells** comprise approximately 20% to 30% of the epididymal epithelium. Basal cells are nonciliated and interdigitate with the principal cells of the epididymal epithelium. Basal cells contain few organelles and their function is unclear. These cells may function in the absorption of fluid from the epididymal lumen. **Narrow cells** are tall columnar cells found in the initial segment of the epididymal duct. The function of these cells is unknown, but they may be involved in endocytosis. **Clear cells** are the precursor to narrow cells and are found in all regions of the epididymal epithelium except the initial segment. Although the function of clear cells is unclear, these cells may be involved in endocytosis. **Halo cells** (lymphocytes and macrophages) comprise 15% to 20% of the epididymal epithelium and are found in all segments of the epididymis. These cells appear to be involved in the immune process.

Mechanism of Sperm Transport and Composition of Epididymal Fluid

The transport time of spermatozoa through the epididymis is species specific. In the rat, sperm transport through the epididymis requires approximately 10 days (corpus - 2 days to 4 days: cauda -5 days to 6 days). Transport of human sperm through the epididymis requires 3 days to 5 days. Peristaltic contraction, ciliary action, difference in hydrostatic pressure along the length of the epididymis as well as hormonal and neural stimulation are all involved in epididymal sperm transport. In some species (e.g. hamster), the hydrostatic pressure is high in the caput epididymis and steadily decreases to the cauda epididymis. This difference in hydrostatic pressure contributes to the transport of spermatozoa from the caput epididymis to the corpus epididymis.

The composition of epididymal fluid changes along the length of the epididymal duct (Figure 3.21). In

FUNCTIONAL ANATOMY OF THE MAMMALIAN MALE REPRODUCTIVE SYSTEM

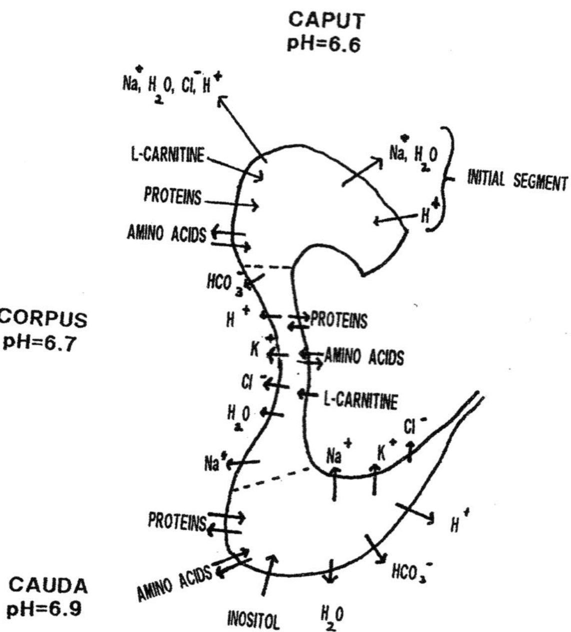

Figure 3.21. Changes in the composition of epididymal fluid.

most mammalian species, approximately 90% of the testicular fluid is absorbed in the efferent ducts. This increases the sperm concentration by a factor of 8. Fluid absorption in the efferent ducts and the initial segment of the epididymal duct is under the influence of estrogen. Hydrogen ion concentration of the epididymal luminal fluid changes as it traverses the epididymis. Luminal fluid entering the caput epididymis has a pH of approximately 7.4. In the caput epididymis, the pH of the epididymal fluid decreases from 7.4 to approximately 6.4, increasing to pH 6.9 as the fluid leaves the cauda epididymis. The acidification of the epididymal fluid in the caput epididymis is important in the maturation of sperm cells. There are also changes in the concentrations of solutes as the epididymal fluid passes through the epididymis. Ionic potassium concentration rises as the fluid moves from the caput epididymis to the cauda epididymis. Ionic calcium concentration decreases as fluid moves from the caput epididymis to the cauda epididymis. Ionic chlorine concentration remains constant in the epididymal fluid along the entire length of the epididymis.

The osmolarity of epididymal luminal fluid varies along the length of the epididymal duct. Rete testis fluid is isoosmolar to blood plasma. Luminal fluid becomes hyperosmolar as it passes through the epididymis. Only 32% and 18% of the osmolarity of the caput epididymal fluid and the caudal epididymal fluid, respectively, can be accounted for by electrolytes. Of the remaining osmolarity, 30% is mostly due to glycerylphosphocholine and carnitine. The remaining substances contributing to the osmolarity of the epididymal luminal fluid is unknown.

<u>Sperm Maturation in the Epididymis</u>

Spermatozoa that leave the testes are not motile. The acquisition of motility can be observed as

FUNCTIONAL ANATOMY OF THE MAMMALIAN MALE REPRODUCTIVE SYSTEM

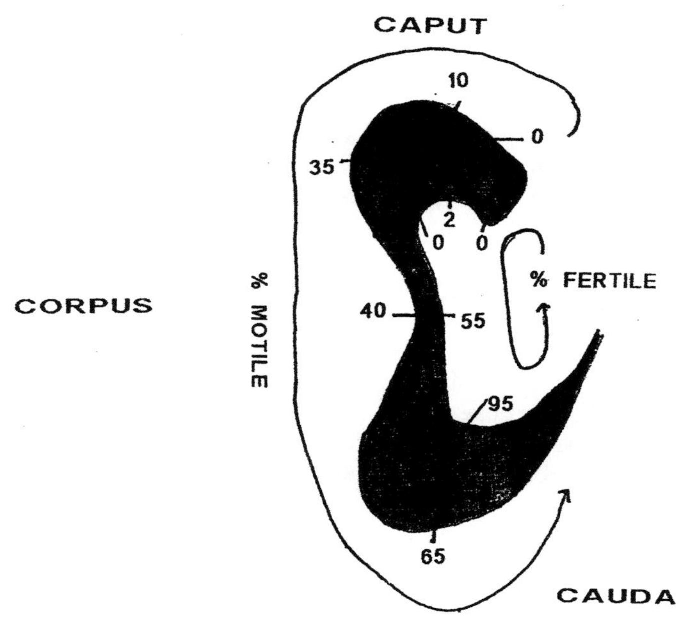

Figure 3.22. The acquisition of sperm motility and fertility along the length of the epididymis.

spermatozoa traverse the epididymis (Figure 3.22). Changes in sperm motility along the length of the epididymal duct can be easily observed in the rat, rabbit and quinea pig, but is less pronounced in humans. In the caput epididymis, the sperm motility increases from 0% to approximately 10%. Motile sperm in the caput epididymis often display a slight vibratory motion. As spermatozoa traverse the corpus epididymis, 40% of spermatozoa show motility. Sperm in the corpus epididymis most often display circular motion of the tail. By the time the spermatozoa have reached the cauda epididymis, 65% of all spermatozoa are motile. Cauda epididymal sperm are capable of unilateral progression with longitudinal rotation of the head. The acquisition of sperm motility appears to be a function of age rather than epididymal microenvironment. Ligation at the distal end of the corpus epididymis in the rabbit results in sperm cells becoming strongly motile even through they are infertile. This shows that motility is not indicative of fertility.

Spermatozoa that leave the testes are not fertile. The acquisition of sperm fertility appears to be a function of exposure of sperm cells to the epididymal microenvironment as they traverse the epididymis. Spermatozoa undergo numerous structural and biochemical changes during their passage through the epididymis. Most of these changes occur in the head and mid-piece region of the cell. Changes that take place in the head of spermatozoa include alteration in the size and shape of the acrosome. These changes are pronounced in many species, but are less pronounced in humans. Disulfide bonds between cystine rich protamines within the chromatin increase nuclear stability. Sperm cell membranes acquire glycoproteins that allow them to attach to the zona pellucida of the ovum during fertilization. Membrane stabilization is increased by disulfide bond formation. The cytoplasmic droplet that formed on the sperm cell during spermiogenesis migrates from the neck down to the mid-piece of the cell and is eventually lost prior to ejaculation. Spermatozoa from the caput epididymis are essentially not fertile. The acquisition of fertility

FUNCTIONAL ANATOMY OF THE MAMMALIAN MALE REPRODUCTIVE SYSTEM

increases dramatically as sperm move from the body to the tail of the epididymis (Figure 3.22). Approximately 95% of the spermatozoa leaving the epididymis are fertile.

<u>The Role of Hormones in Epididymal Sperm Maturation</u>

The ability of the epididymis to maintain sperm viability and to permit maturation is dependent on androgen stimulation. Hypophysectomy or treatment of animals with antiandrogens causes a rapid loss in fertilizing ability of spermatozoa in the cauda epididymis. Testosterone and 5 alpha–dihydrotestosterone exert their effects by maintaining the secretory ability of the epididymal epithelium.

Androgens from the blood and rete testis fluid are responsible for maintaining epididymal function. Plasma testosterone is transported to the epididymis bound to **sex hormone binding globulin**. Epididymal cells convert (by 5 alpha reductase) testosterone to 5 alpha-dihydrotestosterone. Androgens in the rete testis are transported to the epididymis bound to androgen binding protein. This allows the fluid of the rete testis to have a higher androgen level than that of blood plasma and provides a good source of androgens for the caput epididymis. It is not clear whether androgens of the blood or those androgens found in the luminal fluid coming from the testes are responsible for maintaining sperm viability within the epididymis.

Maintenance of the initial segment of the epididymis requires testicular factors. The identification of most of these testicular factors is currently unknown. Fibroblast growth factor secreted by the rete testis has been shown to stimulate secretion of epididymal cells in the initial segments of the epididymis.

FUNCTIONAL ANATOMY OF THE SPERMATIC CORD AND DUCTUS DEFERENS

<u>Anatomy of the Spermatic Cord</u>

The **spermatic cord** extends from the testes to the deep inquinal ring and contains blood vessels, nerves and the ductus deferens. The testicular artery that is outside the abdominal cavity is coiled and forms a vascular cone in the spermatic cord. The **pampiniform plexus** surrounds the coiled artery. The pampiniform plexus is a network of multiple small veins arising from veins outside the testes. The veins reunite forming a single vein near the **inquinal canal**. The pampiniform plexus is involved in testicular thermoregulation by functioning as a heat exchanger. The vascular system of the spermatic cord also eliminates the pulse pressure of the arterial blood.

<u>Anatomy of the Ductus Deferens</u>

The **ductus deferens** is a duct that transports spermatozoa from the epididymis to the urethra (Figure 3.23). It is surrounded by three layers of smooth muscle (inner longitudinal, middle circular and outer longitudinal) that is richly innervated by sympathetic nerves. Sympathetic stimulation produces intense contractions during ejaculation that propels spermatozoa from the ductus deferens into the urethra.

The epithelium of the ductus deferens is pseudostratified and the lamina propria is thrown up into longitudinal folds. These folds permit expansion of the duct during ejaculation. The distal end of the duct is dilated and receives the duct draining the seminal vesicles thus forming a short **ejaculatory duct**. The

FUNCTIONAL ANATOMY OF THE MAMMALIAN MALE REPRODUCTIVE SYSTEM

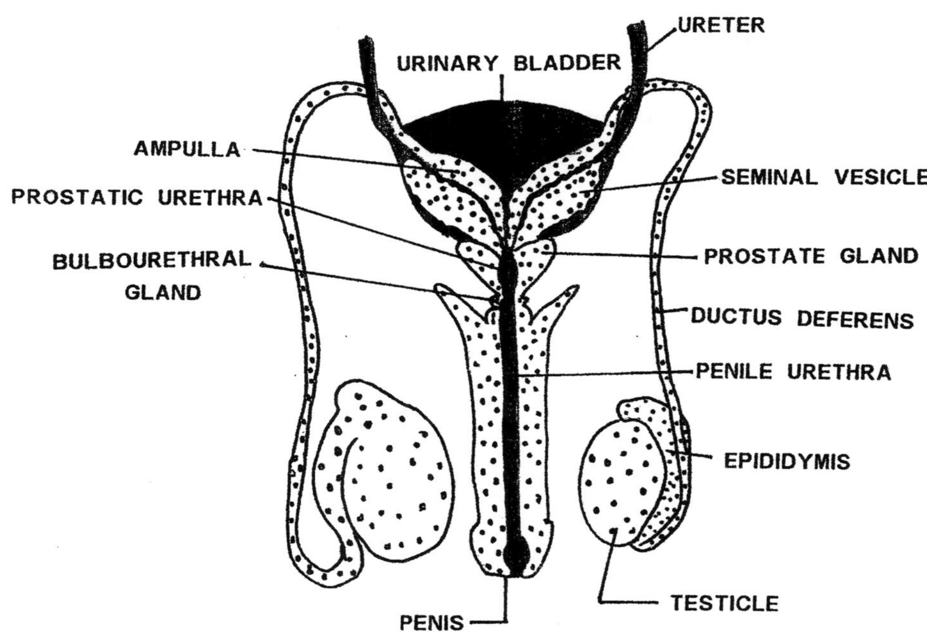

Figure 3.23. Anatomy of the mammalian male reproductive system.

ejaculatory ducts arising from each gonad converge to join the urethra as it passes through the prostate gland.

FUNCTIONAL ANATOMY OF THE ACCESSORY SEX GLANDS AND THE COMPOSITION OF SEMEN

Anatomy of the Seminal Vesicles

The **seminal vesicle** is a contorted, branched, saccular diverticulum of each ductus deferens. Its excretory duct units to form the ejaculatory duct (Figure 3.23). The central lumen of the seminal vesicle is irregular and recessed giving it a honey combed appearance. The seminal vesicle is composed of two layers of smooth muscle (inner circular and outer longitudinal) that is richly innervated by sympathetic nerves. The epithelium is pseudostratified and consists of secretory cells that produce a yellowish viscous fluid

FUNCTIONAL ANATOMY OF THE MAMMALIAN MALE REPRODUCTIVE SYSTEM

primarily composed of fructose, fibrinogen and vitamin C. This fluid contributes to the nutritive and supporting component of semen.

Anatomy of the Prostate

The **prostate** is a large gland that surrounds the neck of the urinary bladder and the superior part of the urethra (Figure 3.23). It is divided into glandular lobules whose ducts empty into the **prostatic urethra**. The epithelial lining varies from inactive low cuboidal to active columnar depending on the degree of androgen stimulation. The stroma and the capsule contain dense fibrous connective tissue and numerous smooth muscle fibers. The smooth muscle is innervated by the sympathetic nervous system. Sympathetic stimulation causes powerful contractions during the ejaculatory process. Prostatic secretion is milky in appearance and contains citric acid and a variety of hydrolytic enzymes. This fluid liquefies semen after deposition in the female reproductive tract.

Anatomy of the Bulbourethral Gland

The **bulbourethral (Cowper's) gland** (William Cowper, English anatomist, 1666 – 1709) is associated with the caudal portion of the urethra and drains into the pelvic urethra at the bulbar region (Figure 3.23). The Cowper's gland of bulls, rams and stallions is relatively small. The gland is totally absent in the dog. The Cowper's gland of the boar is large and produces a gel-like component found in the ejaculate.

Semen Composition

Semen is composed of sperm cells and a liquid portion called **seminal plasma**. Sperm cells are of testicular origin, whereas most of the seminal plasma is a product of the accessory sex glands. The seminal plasma contains a number of substances that occurs in higher concentrations when compared to the blood. These substances include fructose, citric acid, inositol, ergothioneine, glycerylphosphocholine, glutamic acid and certain enzymes. The concentration of these substances in the seminal plasma is dependent on the amount of testosterone that is secreted by the testes. The concentration of these substances in the seminal plasma can also vary from one ejaculate to another from the same animal.

The **ejaculate** of many species can be divided into three fractions: the presperm fraction, the sperm rich fraction, and the postsperm fraction. The **presperm fraction** is watery and contains ergothioneine and citric acid. This fraction has few sperm cells. The **sperm rich fraction** has a high concentration of sperm cells and also contains ergothioneine that primarily comes from the ampullary region of the ductus deferens. The **post sperm fraction** is composed of secretions from the seminal vesicles. This fraction contains few sperm cells but is rich in citric acid.

MAMMALIAN INTROMITTENT ORGANS

Functional Anatomy of the Mammalian Penis

The mammalian **penis** is an intromittent organ for the deposition of semen in the female reproductive tract. Penile characteristics vary considerably between mammalian species. The erect penis of whales (e.g. Blue Whale and Humpback Whale) can be 8 feet long and 1 inch in diameter. Bull elephants have an erect penis approximately 5 feet long. The erect penis of the boar is approximately 18 inches long. The gorilla and the orangutan each have an erect penis that is less than 2 inches long. The erect penis of a chimpanzee is approximately 3 inches long when erect. The erect penis of the adult human male is the longest, thickest and most flexible penis of any living primate averaging 5 inches in length (longest human penis medically recorded is 13 inches!). The free end of the mammalian penis can also vary in shape (Figure 3.24). Males of mammalian species that produce a hard copulatory plug that binds to the lining of the female reproductive tract often have a thin pointed penis with backward pointing spines or barbs. These penises are adapted for the removal of hard plugs. Penises that are straight and piston in shape with an acorn shape end (e.g. humans and stallions) are designed to remove soft or liquid coagulum from the female reproductive tract.

The penis of several mammalian species (including most primates) contains a bone called the **baculum** (also called the **os penis**) (Figure 3.25). The baculum of many mammalian species (including many primates) is forked. A few primates (e.g. lemur) have a baculum that is straight or slightly curved. The baculum can vary considerably in length. Penile erection in primates containing a baculum involves muscular control similar to a winch raising a rigid structure. The penis of ruminants and swine is a **sigmoid flexure**, which is an S shaped penis that can be withdrawn completely into the prepuce (Figure 3.26). Movement of the sigmoid flexure penis is dependent on a pair of retractor penis muscles attached to the ventral side of the penis just anterior to the sigmoid flexure. Relaxation of the muscles allows extension of the penis, whereas contraction of the muscles draws the penis back into the prepuce.

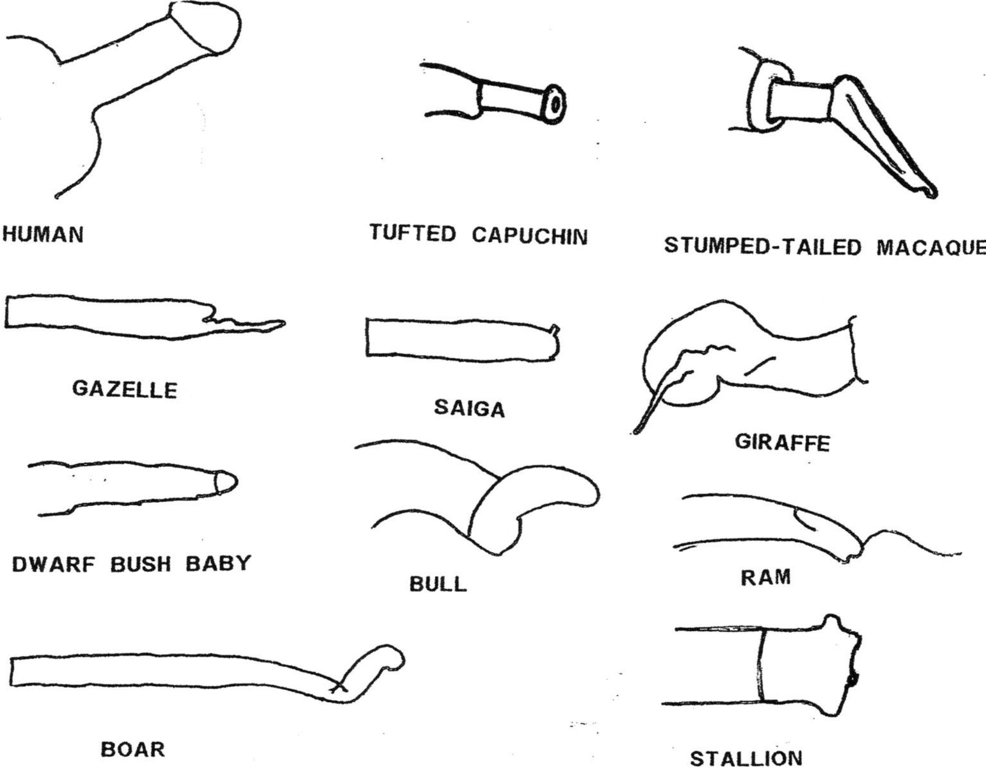

Figure 3.24. Variation in shape of the free end of the penis in some mammalian species.

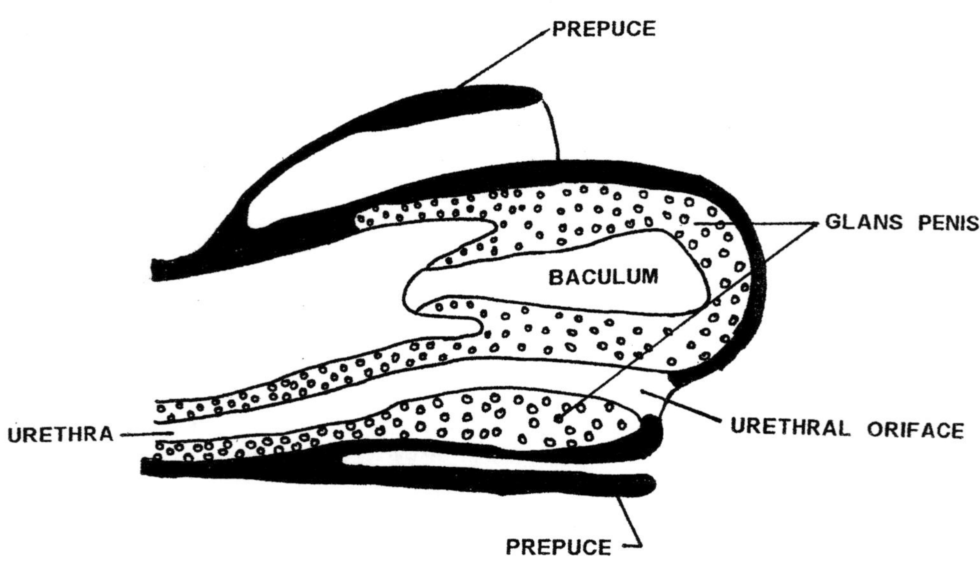

Figure 3.25. The baculum of a primate penis.

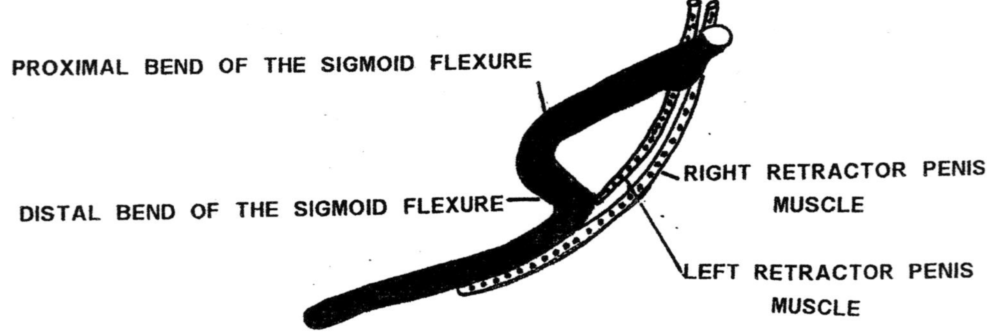

Figure 3.26. Anatomy of a sigmoid flexure penis. Lateral view showing retractor muscles of the penis in the bull, ram and boar.

FUNCTIONAL ANATOMY OF THE MAMALIAN MALE REPRODUCTIVE SYSTEM

A few mammalian species (e.g. humans and stallions) have an **erectile penis** that contains no baculum. The human penis is composed of the root, body, and glans (Figure 3.27). The **body of the penis** is composed of three cylindrical masses. The **corpus spongiosum** is erectile tissue located at the midventral region of the penis and surrounds the urethra (Figure 3.28). The corpus spongiosum is enlarged at the ischial arch and forms the bulb of the penis. The corpus spongiosum contains cavernous spaces for blood accumulation, but expansion of these spaces is limited by the tunica albuginea. The **corpora cavernosa** occurs as paired bodies and is located in the dorsilaterial regions of the penis (Figure 3.28). The tunica albuginea covers and invades these bodies forming trabeculae that help support this cavernous tissue. The corpora cavernosa of humans and stallions contains large cavernous spaces for the accumulation of blood during erection. In ruminants and swine these spaces are smaller except at the cura and the distal end of the penis. The **root of the penis** is composed of the bulb and crux. The bulb of the penis is the expanded portion of the corpus spongiosum and is attached to the urogenital diaphragm. The crux of the penis is the tapered portion of the corpora cavernosa. The **glans** is the most distal part of the penis and is composed of several parts (Figure 3.27). The **glans penis** is the enlarged end of the corpus spongiosum and swells during erection. The **corona glandis** is the margin of the glans penis. The **prepuce** is the foreskin that covers the glans penis. This sheath is often removed in humans in a process known as **circumcision**. Arteries supplying the penis dilate and large volumes of blood enter the sinuses during penile erection. Expansion of the sinuses compresses venous drainage so blood is retained causing penile erection.

<u>Evolution of the Erectile Human Penis</u>

Erection of the human penis depends on an unusual system of vasocongestion. Loss of the penile baculum gives the human penis greater flexibility for a greater range of copulatory movements. The rhythmicity of copulatory thrusts seems designed to maximize the intensity of tactile stimulation delivered to the female genitals. The human penis may have became flexible and streamline because ancestral females favored whole body movements and intense tactile stimulation of their genitals. The human penis may have evolved as a tactile stimulator for use in copulatory courtship. In humans, hidden ovulation and continuous sexual receptivity can give the female an excellent opportunity to test a male sexual partner for endurance while running a low risk, per copulation, of an unwanted pregnancy. Whole body movements require more energy and could have been used as a possible fitness indicator of the male.

THE MALE SEXUAL RESPONSE CYCLE

<u>General</u>

Many thoughts and stimuli can be sexually arousing. When adequate stimuli and thoughts are present, males become sexually aroused and will display a sexual response cycle. Although there are slight variations in the sexual response cycle between and within species, the cycle can be generally divided into four phases: the excitement phase, the plateau phase, the orgasmic phase and the resolution phase. The sexual response cycle for the human male is discussed below.

<u>Excitement Phase</u>

In the **excitement phase** of the male sexual response cycle penile erection occurs in response to sexual arousal. Erotic stimuli of visual, olfactory and auditory sensation are detected by receptors and converted to bioelectrical information that is transmitted to the erection center located in the upper regions of the spinal cord. Erotic thoughts of the cerebrum are also transmitted to the spinal erection

FUNCTIONAL ANATOMY OF THE MAMMALIAN MALE REPRODUCTIVE SYSTEM

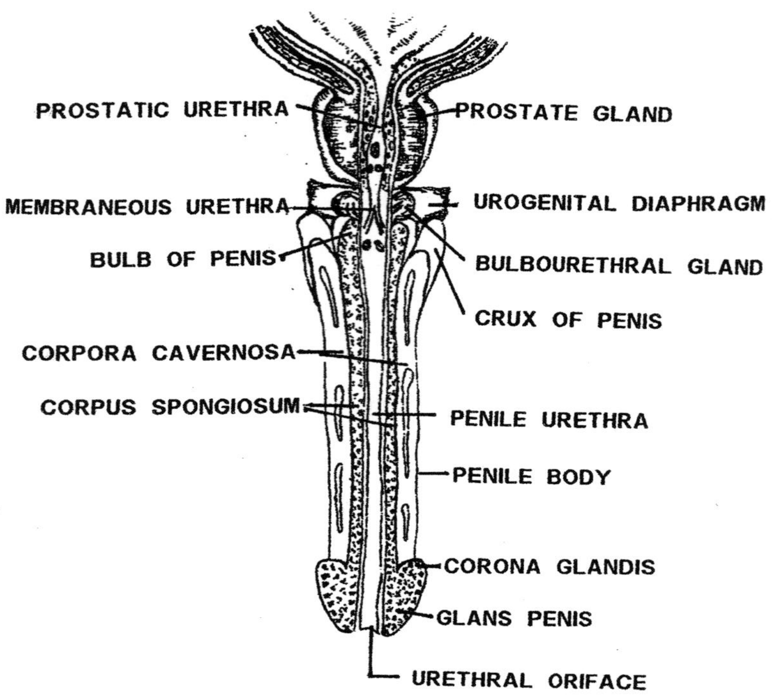

Figure 3.27. Longitudinal view of the erectile penis.

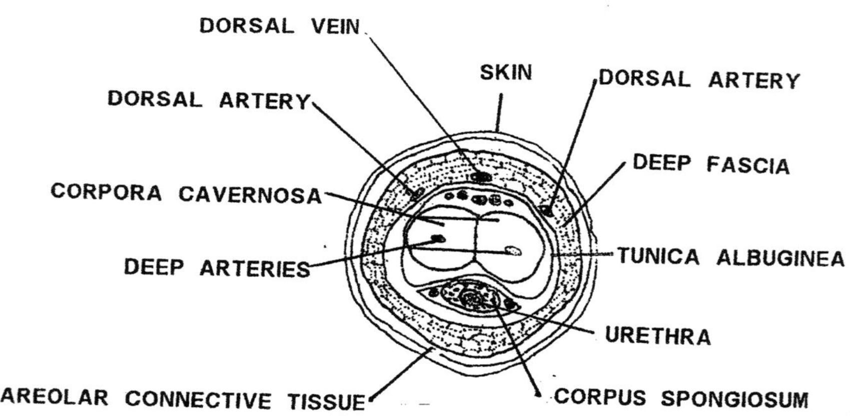

Figure 3.28. Transverse view of the anatomy of an erectile penis.

FUNCTIONAL ANATOMY OF THE MAMMALIAN MALE REPRODUCTIVE SYSTEM

center. From the spinal erection enter, bioelectrical information is carried along parasympathetic nerves to the blood vessels of the penis. These nerves release neurotransmitter acetylcholine and vasoactive intestinal peptide that cause dilation of the penile arterioles. This results in vasocongestion of the erectile tissue causing penile erection. Since blood flow through the arterioles is greater than that leaving the penis, the erection is maintained. Bioelectical information from the cerebrum (e.g. in the form of fear and stress) can also inhibit penile erection. Since erection is mostly a spinal cord reflex, men who have their spinal cords severed above the site of the erection center still have an erection even though they cannot feel it.

Other physiological sensations can also be noted during the excitement phase. There is a widening of the urethral opening. The scrotal skin becomes congested and thick thereby reducing the scrotal diameter. Contraction of the cremaster muscle elevates the testes within the scrotum. Nipples of the breast become erect. Areas of the skin become reddened due to vascular dilation. There is an increase in blood pressure, heart rate and the depth and rate of pulmonary ventilation. There is an increase in tension of skeletal and smooth muscles throughout the body.

Plateau Phase

The **plateau phase** can be best described as a continuation of the excitement phase. Penile erection continues. There is a slight increase in the size of the glans penis and its color darkens. There is a swelling of the corona glandis. There is often preorgasmic emission of a few drops of semen, which may contain a few sperm. The testes become more elevated in the scrotum and slightly rotated. The volume of the testes may increase by as much as 50% due to the accumulation of fluid. There is a further increase in heart rate, blood pressure as well as the depth and rate of pulmonary ventilation. There is also a further increase in tension of skeletal and smooth muscles.

Orgasmic Phase

The male may enter the **orgasmic phase** many times in a few minutes. There is a great release of neuromuscular tension. Clutching and clawing motion of the hands and feet can be observed. The testes are at the maximal elevation within the scrotum. The cardiac rate peaks at about 180 beats per minute and the blood pressure averages about 200/100. The respiratory rate peaks at about 41 breaths per minute. The sexual flush peaks in intensity and duration.

Ejaculation is the expulsion of semen and is controlled by the ejaculation reflex. The ejaculatory reflex center of the spinal cord is located superior to the erection center. When this center is activated, it sends sympathetic neural stimulation to the muscles at the base of the penis. Ejaculation occurs in two stages called the emission phase and the expulsion phase. In the **emission phase of ejaculation** there is a consecutive sequence of contractions of the smooth muscles in the wall of the testes, epididymides, vas deferens, ejaculatory duct, seminal vesicles, prostate, Cowper's gland and urethra. Simultaneously, the sphincter that guards the opening of the urinary bladder contracts preventing urine from entering the semen and semen from entering the bladder. The **expulsion stage of ejaculation** begins with rhythmic contractions of the bulbocavernosus muscle, which lies at the base of the penis. The first three to four contractions are intense and result in a forceful expulsion of most of the semen in the urethra. The following contractions are less intense and produce small squirts of semen from the urethra. The urethral contractions are generally 0.8 seconds apart.

Immediately after ejaculation, the male enters into a **refractory period** in which erotic stimuli are not effective in causing and maintaining erection of the penis until the sexual tension has decreased to near resting levels. The length of this period can vary within a male and between males. Generally, the refractive period may last several minutes in younger males whereas an hour may be required in older males.

<u>Resolution Phase</u>

The **resolution phase** can be best described as the arousal mechanisms returning to their resting levels. The resolution phase occurs during and after the refractory period. Almost immediately, the penis looses its erection. Sympathetic stimulation from the spinal erection center causes the arterioles of the penis to constrict, thus reducing vasocongestion of the erectile tissue. Other immediate changes include lowering of the cardiac rate, depth and rate of pulmonary ventilation, and blood pressure. Physiological changes that take a few minutes to several hours include final reduction in penis size, relaxation of the scrotum, descent of the testes and loss of nipple erection. About 30% of all men sweat over their body. Some men may have an intense desire to sleep.

FUNCTIONAL ANATOMY OF THE MAMMALIAN FEMALE REPRODUCTIVE SYSTEM

THE OVARY

Ovarian Location and Structure

The **ovaries** are a pair of bodies located one on each side of the uterus (Figure 4.1). The shape of the ovary varies between species. Most ovaries have shapes resembling almonds, berries or kidneys. Each ovary can be divided into the cortex and medulla. The **ovarian medulla** is composed of irregular fibroelastic connective tissue with extensive blood vessel and nerve innervations. The **ovarian cortex** contains many fibroblasts, protein fibers and smooth muscle tissue. The tunica albuginea is connective tissue cells arranged in a parallel fashion near the cortical surface. Numerous **follicles** that contain the female gametes in various stages of development are found on the cortical surface.

Follicular Anatomy and Development

Human follicular development (Figure 4.2) begins at approximately 18 weeks to 20 weeks of gestation. Follicular formation involves the separation of the oocyte from the ovarian stroma, the association of **granulosa cells** (derived from the ovarian cortex) around the surface of each oocyte, and the formation of the basement membrane. **Primordial follicles** are oocytes surrounded by flattened granulosa cells. The diameter of primordial follicles is approximately 50 micrometers. Receptors for FSH are expressed on the granulosa cells of the primordial follicles. Granulosa cells and the oocyte communicate with each other during follicular development. Low molecular weight sugars, amino acids, lipid precursors and nucleotides are transferred between the oocyte and granulosa cells by gap junctions. The oocyte secretes proliferative factors (e.g. growth differentiation factor 9 and bone morphogenic protein) that influence granulosa cell division. Vascular channels from the ovarian medulla penetrate the developing follicle as follicular development proceeds.

Stimulation of the primordial follicles by FSH causes the granulosa cells to become cuboidal in shape thus forming **primary follicles**. Formation of human primary follicles from primordial follicles takes approximately 150 days. The granulosa cells of each primary follicle undergo mitosis forming multiple layers of cells around the oocyte. This structure is called the **secondary follicle**. Approximately 120 days are required for human primary follicles to develop into secondary preantral follicles (the development of a primordial follicle into a secondary preantral follicle requires approximately 9 menstrual cycles!). All stages of follicular development from primordial follicles to the preantral secondary follicles are observed on the ovaries of the female human neonate as early as 4 months to 6 months of age.

 Secondary follicles that do not become atretic develop into **tertiary follicles** (also called Graffian follicles, [Reijnier de Graff, Dutch physiologist and histologist, 1641 – 1673]). The development of a human tertiary follicle from a cohort of preantral secondary follicles occurs during the follicular phase of the menstrual cycle. Follicle stimulating hormone receptor signaling of the dominant follicle causes it to develop faster than other cohort follicles. The dominant follicle undergoes increased follicular fluid formation, granulosa cell proliferation and increased expression of LH receptors on granulosa cells. The dominant follicle produces higher levels of estradiol 17-*B* and inhibin-B. Estradiol 17-*B* and inhibin-B depress the secretion of pituitary FSH during the midfollicular phase of the menstrual cycle. Cohort secondary preantral follicles

FUNCTIONAL ANATOMY OF THE MAMMALIAN FEMALE REPRODUCTIVE SYSTEM

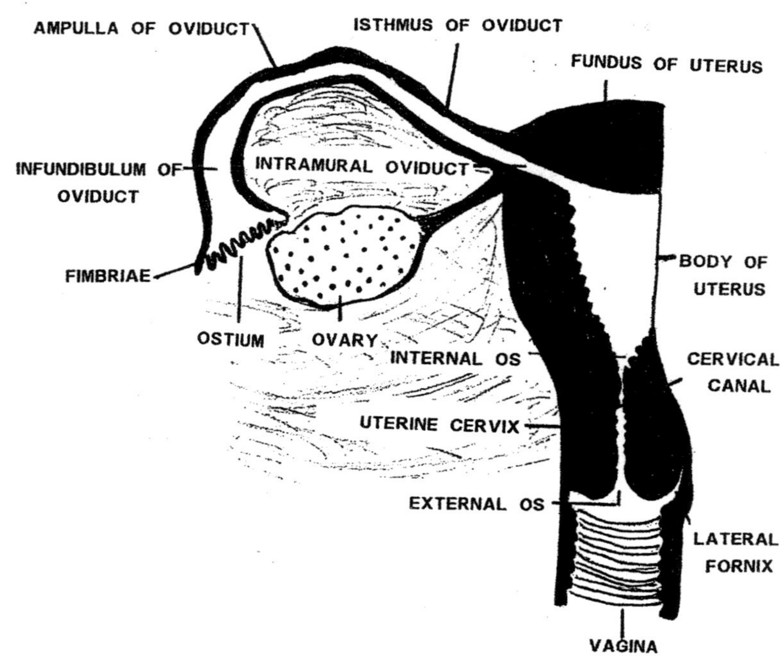

Figure 4.1. Anatomy of the human female reproductive system.

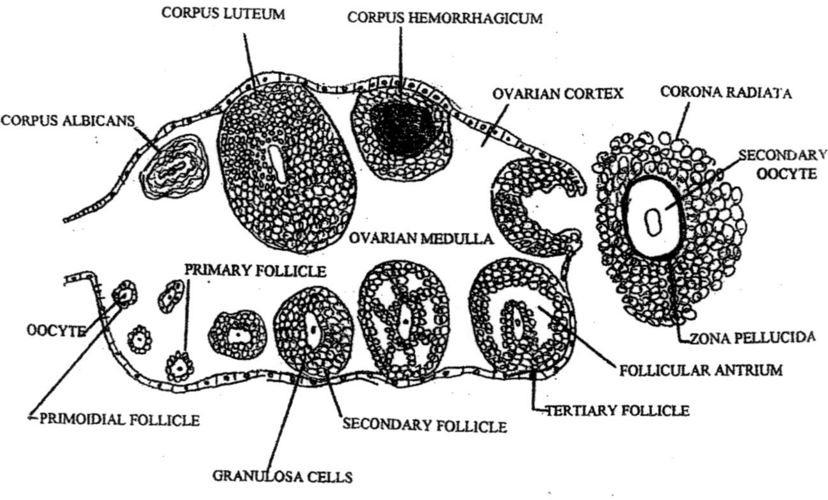

Figure 4.2. The ovarian cycle.

FUNCTIONAL ANATOMY OF THE MAMMALIAN FEMALE REPRODUCTIVE SYSTEM

deprived of adequate FSH stimulation undergo massive apoptosis of their oocyte and granulosa cells. The granulosa cells of tertiary follicles secrete fluid that starts to accumulate in the intercellular spaces of each follicle. Blood filtrate also contributes fluid to the growing **follicular antrum**. Several layers of cells surrounding the dominant follicle differentiate into the **theca interna** and the **theca externa**. Cells of the theca interna are glandular in appearance and the layer is highly vascularized. The theca externa contains dense connective tissue and is less vascularized when compared to the theca interna. Development of human tertiary follicles from secondary preantral follicles requires approximately 15 days.

<u>Oogenesis</u>

Oogenesis is the process of formation and development of the ovum. Human oogenesis is observed on the ovary as early as the 11th to 12th week of gestation. **Gonocytes** (primordial germ cells) migrate to the genital ridge and differentiate into **oogonia**. Oogonia undergo a finite number of mitotic divisions (Figure 4.3). Near the time of birth in most mammalian species, the oogonia complete their mitotic divisions and enter into meiosis and become arrested in the diplotene stage of prophase I. These meiotically arrested cells are referred to as **primary oocytes**.

Luteinizing hormone reinitiates oogenesis during each menstrual cycle of the human. Follicular granulosa cells have cytoplasmic processes that form tight junctional complexes with the developing oocyte membrane (Figure 4.4). The granulosa cells synthesize a glycocalyx that surrounds the developing oocyte. When the cellular processes of the granulosa cells retract, a matrix known as the **zona pellucida** surrounds the developing oocyte. The zona pellucida is composed of a family of glycoproteins referred to as **zona pellucida (ZP) 1, ZP2, and ZP3**. Completion of the first meiotic division produces two haploid eggs within the follicle. This reduction division is unequal producing a large haploid **secondary oocyte** and a small haploid polar body (often referred to as the **first polar body**). The secondary oocyte (still in the follicle) begins the second meiotic division only to become arrested at metaphase II. The secondary oocyte is then ovulated at this stage of development. Sperm penetration at the time of fertilization reinitiates meiosis of the secondary oocyte. This results in the unequal division producing a large **ootid** (pronucleated egg), that becomes fertilized and a small **second polar body**.

Why does human oogenesis arrest in the diplotene stage of prophase I? Accumulation of deleterious mutations in mitochondrial (mt) DNA during an organism's lifetime appears to be a contributing factor to the aging process. The rate of mutation of mt-DNA is correlated with cellular activity. This mutation is due to the production of peroxides and oxygen free radicals that damage organic molecules in close proximity of mitochondria. A prolong period of relative metabolic inactivity of the primary oocyte may be a protective mechanism for inhibiting the mutation of the egg's mt-DNA. Since the mt-DNA lacks mechanisms of repair, it is important to ovulate secondary oocytes that will contribute healthy mt-DNA to the future embryo.

<u>Ovulation of the Secondary Oocyte</u>

Ovulation is the release of the secondary oocyte from the ovarian follicle. At the time of ovulation, a small avascular region known as the **stigma** appears on the surface of the tertiary follicle. Pressure within the follicle causes the follicular membrane to tear at the stigma. The free-floating secondary oocyte surrounded by the **corona radiata** and the antral fluid leaves the follicle through the tear in the wall.

Ovulation in the mammalian female is a complex phenomenon occurring in response to biophysical, biochemical and physiological mechanisms (Figure 4.5). The process of ovulation is still not completely understood. Ovulation in most domestic mammalian species (except the mare) occurs approximately 38

55

FUNCTIONAL ANATOMY OF THE MAMMALIAN FEMALE REPRODUCTIVE SYSTEM

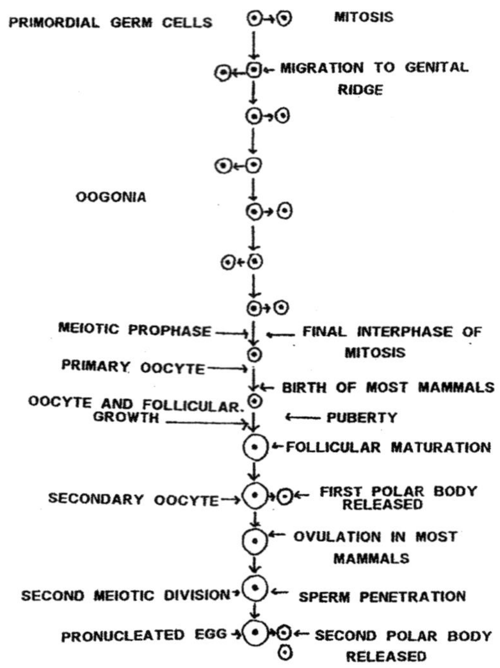

Figure 4.3. Mammalian oogenesis.

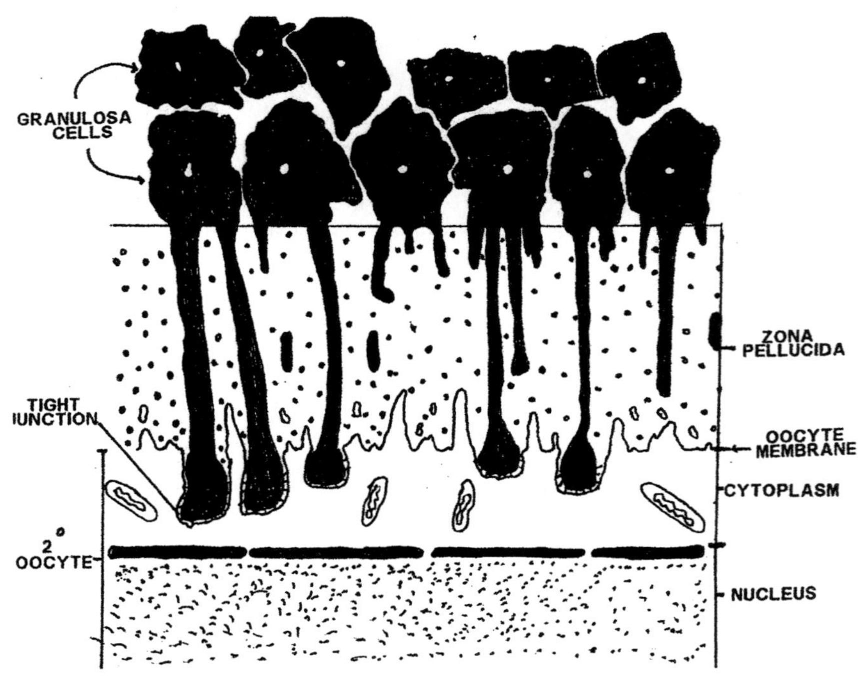

Figure 4.4. Anatomy of the follicular secondary oocyte and the surrounding cumulus granulosa cells

THE FUNCTIONAL ANATOMY OF THE MAMMALIAN FEMALE REPRODUCTIVE SYSTEM

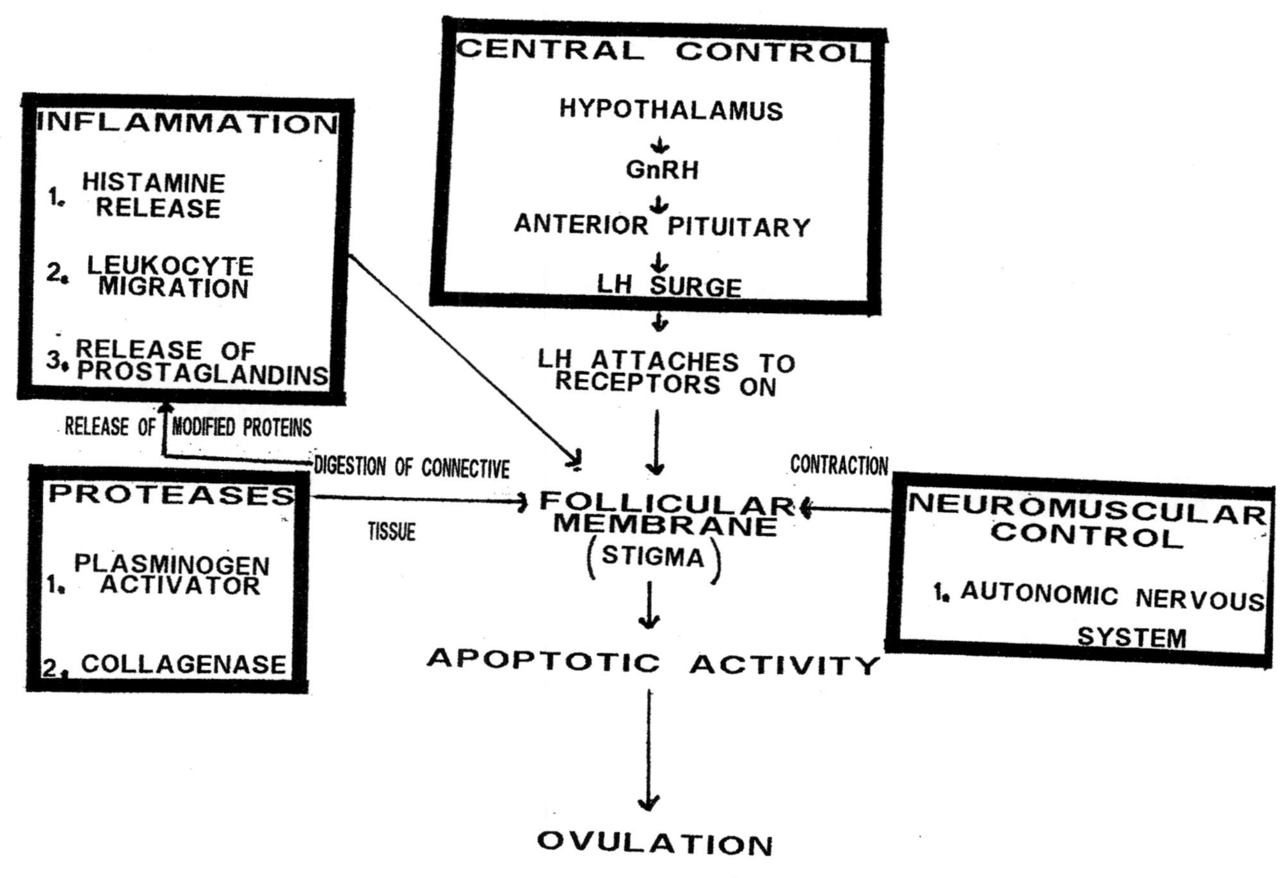

Figure 4.5. Possible mechanism involved in ovulation in the mammalian female.

hours to 42 hours after the onset of the LH surge. Gonadotropin stimulation rapidly causes (1) increased release of ovarian prostaglandins (2) increase ovarian steroidogenesis (3) release of ovarian plasminogen activator. Each of these events contributes to apoptotic activity in the area of the stigma and eventual ovulation of the secondary follicle. Each event is discussed below.

The LH surge results in the stimulation of ovarian prostaglandin synthesis ($PGF_{2\,alpha}$ and PGE_2). The site of the production of ovarian prostaglandins is currently unknown. The role of prostaglandins in the ovulatory process currently is not completely understood. Prostaglandin F_{2alpha} causes vasoconstriction of follicular blood vessels in the area of the stigma contributing to cellular apoptosis. Prostaglandin E_2 has been shown to be important in remodeling of follicular cells after ovulation

FUNCTIONAL ANATOMY OF THE MAMMALIAN FEMALE REPRODUCTIVE CYCLE

resulting in the formation of the corpus luteum.

The preovulatory LH surge rapidly causes increased steroidogenesis in the cells of the theca interna. Elevated levels of ovarian steroids activate and proliferate fibroblasts under the theca interna. Fibroblasts produce **collagenase** that is responsible for collagenolysis of the follicular wall in the area of the stigma. Collagenolysis of the follicular wall releases modified proteins that initiate an acute inflammation. Acute inflammation of the follicular wall includes the release of histamine, leukocyte migration to the inflammatory area and the further release of prostaglandins. This inflammatory reaction in the stigma of the follicle contributes to cellular apoptosis and ovulation.

The preovulatory surge of LH causes the mural granulosa cells and the cumulus granulosa cells to produce **plasminogen activator**. Cumulus granulosa cells also produce hyaluronic acid that is responsible for attracting water into the follicle thus increasing the volume of the antral cavity. Plasminogen activator is responsible for catalyzing the reaction of **plasminogen** to **plasmin.** Plasmin is involved in the disassociation of the connective tissue matrix of collagen fibers at the stigma of the follicular wall. Plasmin is also indirectly involved in the inflammation of the follicular wall by converting procollagenase to collagenase.

Neuromuscular activity may also contribute to the ovulatory process in some mammalian species. The ovarian stroma and the theca externa of the follicle contain smooth muscles cells that are innervated by the autonomic nervous system. After thinning of the follicular stigma, neural stimulation of smooth muscles cells of the ovarian stoma may facilitate rupture of the follicle. Stimulation of the smooth muscle cells of the theca externa by PGF_{2alpha} may contribute to the extrusion of the secondary oocyte from the follicle.

Development of the Corpus Luteum

Once ovulation has occurred, the basement membrane of the follicle disintegrates and the granulosa cells and theca cells start to reorganize. Angiogenesis occurs from preexisting follicular blood vessels and blood fills the old antral cavity forming the **corpus hemorrhagicum.** The remaining granulosa cells begin mitosis and invade the old antral cavity forming the **corpus luteum (CL)** (Figure 4.2). The CL has 2 population of cells. Follicular thecal cells give rise to **small luteal cells** that produce progesterone. Follicular granulosa cells give rise to **large luteal cells**. Large luteal cells produce 10 times to 20 times more progesterone than small luteal cells. Large luteal cells also produce moderate levels of estradiol 17 - B, relaxin, oxytocin and inhibin-A in many mammalian species. Secretion of these hormones in many mammalian species is dependent on LH and prolactin. If fertilization occurs, the corpus luteum survives to produce hormones during the first third of pregnancy. If fertilization does not occur, the corpus luteum degenerates and fills with connective tissue forming the **corpus albicans.**

Ovarian Function and Energy Stress

Mammalian ovarian function varies in a dose response manner to the amount of **energy stress** organisms experience. Follicular estrogen synthesis is reflective of growth and degree of follicular maturation. In mammals that experience an adequate energy balance, larger and more numerous follicles develop, thus producing more estrogen. Mammals that experience energy stress often have **follicular suppression.** Follicular suppression is characterized by smaller and less numerous follicles that have a slower period of development. This results in a shorter ovarian follicular phase and a longer luteal phase. Follicular suppression is also characterized by lower estrogen production.

FUNCTIONAL ANATOMY OF THE MAMMALIAN FEMALE REPRODUCTIVE SYSTEM

Luteal progesterone production is accomplished by the luteinization of follicular granulosa cells and theca cells. Ovarian follicular suppression in response to energy stress often results in **luteal suppression**. Energy stress often leads to smaller follicles primarily due to decreased granulosa cell proliferation. This often leads to smaller CL's because of the decreased number of luteinized granulosa cells. Although luteinized theca cells maintain their response to LH stimulation, luteinized granulosa cells are less in number and less responsive to LH stimulation. This results in the decreased synthesis of progesterone.

Ovarian function can vary depending on the degree of energy stress. **Anovulation** is the suppression or cessation of ovulation. Human anovulatory cycles ending in menstruation are often caused by follicles starting to develop but failing to mature. These follicles produce enough estrogen to achieve endometrial proliferation. Since follicles fail to mature, no CL's develop; plasma progesterone levels are low resulting in early menses. Human anovulatory cycles without menstruation are often caused by suppressed follicular development resulting in extremely low estrogen production. Extremely low levels of estrogen usually result in the development of a thin endometrial lining that is reabsorbed. **Amenorrhea** is the absence or cessation of menses. In human amenorrhea, there is no follicular development, no production of estrogen and no uterine development.

THE OVIDUCTS

<u>Anatomy of the Oviducts</u>

The **oviducts** are paired tubes extending from near each end of the ovary to the top of the uterus. These structures are also called **Fallopian tubes** [Gabriele Fallopius, Italian anatomist, 1523-1562]. The human oviduct has a diameter of a soda straw and is approximately 10 cm in length.

The oviduct can be divided into several regions along its length (Figure 4.1). The **infundibulum** is the funnel shaped portion of the oviduct in close proximity to the ovary. The edges of the infundibulum have finger-like projections called **fimbrae**. The opening of the infundibulum in which the ovum enters is called the **ostium**. Proceeding along the length of the oviduct, the next area is called the **ampulla**. This area is widened and is the site of fertilization in many species of mammals. The narrow segment of the oviduct between the ampulla and the uterus is called the **isthmus**. The **intramural oviduct** is the portion of the oviduct embedded in the uterine wall. The point where the oviducts enter the uterine cavity is called the **uterotubal junction**.

Microscopic examination of the oviduct in cross section reveals three layers. The outer layer is protective connective tissue and is referred to as the serosa. The middle layer is composed of smooth muscle (inner circular layer and an outer longitudinal layer) and is called the muscularis. The inner most layer is the mucosa. This layer contains ciliated and nonciliated cells as well as a few mucous glands. Ciliary movement plays an important role in ovum, embryo and sperm cell transport. Estrogens cause the secretion of mucous, ciliary movement and muscular contractions. Progesterone has the opposite effect.

FUNCTIONAL ANATOMY OF THE MAMMALIAN FEMALE REPRODUCTIVE SYSTEM

THE UTERUS

Gross Anatomy of the Uterus

Uterine anatomy varies among mammalian species. Below is a summary of the uterine types that can be found in various species (Figure 4.6).

Uterus duplex is typically found in many rodent species. The reproductive system of these animals is composed of two separate oviducts. The inferior end of each oviduct expands into an enlarged area referred to as the uterus (there are two uteri in these species). Each uterus empties into the vagina.

Uterus bipartite is typically found in many carnivores. The lower parts of the uteri create a single body divided into two by a partition that represents the two walls of the uteri. The upper parts of the uteri remain separate as horns.

Uterus bicornate is found in many species of unglulates. This uterine type is similar to uterus bipartite except that the separation between the uteri has disappeared.

The last and most advanced uterine type is **uterus simplex**. This uterine type is found among apes and humans. Both vaginae and uterus are fused along their entire length leaving the oviducts separate.

Anatomy of Uterus Simplex

The human uterus is a single inverted pear-shaped structure situated in the pelvic cavity above the urinary bladder and in front of the rectum (Figure 4.1). The uterus is 7.5 cm long, 5 cm wide and 1.75 cm thick in nulliparous women. In multiparous women the uterus is larger and more variable in shape.

The primate uterus is supported in the abdominal cavity by bands and cords of tissue. The **broad ligaments** attach each side of the uterus to the pelvic wall. The **uterosacral ligaments** attach the lower end of the uterus to the tailbone. The **lateral cervical ligaments** attach the cervix to the pelvic wall. The **round ligaments** attach the portion of the uterus near the oviducts to the lower pelvic wall. If a longitudinal section is made through the primate uterus, several distinct regions can be observed. The domed shape region above the uterotubal junctions is called the **fundus**. The **corpus** is the tapering portion, which ends at the external constriction. The narrow region inferior to the corpus is the **cervix**. The **cervical canal** is a small channel within the **cervix** that connects the **vagina** to the uterine cavity. The opening of the cervical canal into the uterine cavity is called the **internal cervical os**, whereas the opening of the cervical canal into the vagina is called the **external cervical os**. The cervix has a diameter of 1 to 2 inches.

Uterine Microanatomy

In humans, the walls of the uterine body and fundus are composed of three layers of tissue (Figure 4.7). The external layer is referred to as the **uterine perimetrium**. This layer is composed of connective tissue that protects the uterus. The middle layer is called the **uterine myometrium**. It is composed of an inner longitudinal, a middle circular and an outer longitudinal layer of smooth muscle tissues. The inner most lining is called the uterine **endometrium**. The endometrium can further be subdivided into an internal layer, the **stratum functionalis**, and a deep layer, the **stratum basilis**. The stratum functionalis of the

FUNCTIONAL ANATOMY OF THE MAMMALIAN FEMALE REPRODUCTIVE SYSTEM

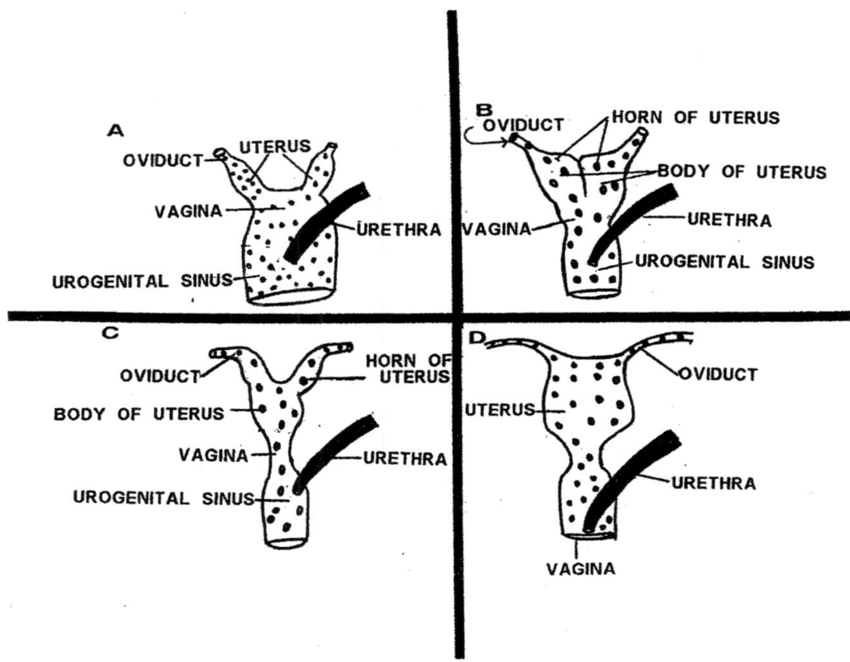

Figure 4.6. Various types of mammalian uteri. (A) Uterus duplex. (B) Uterus bipartite. (C) Uterus bicornate. (D) Uterus simplex.

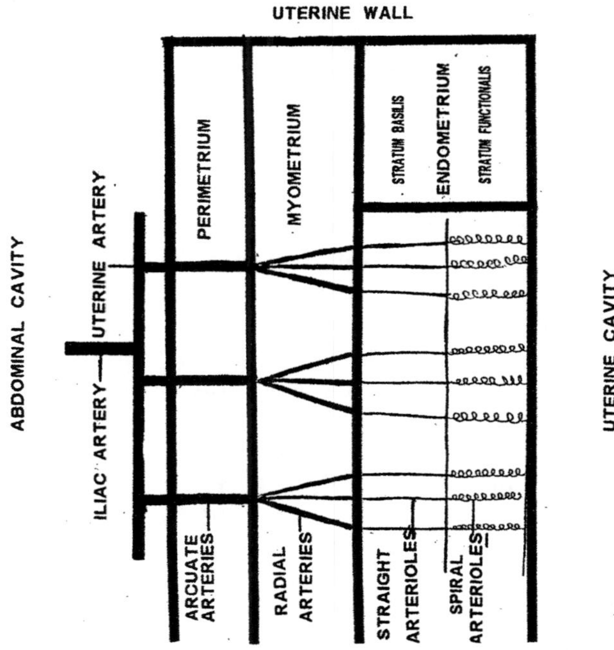

Figure 4.7. Anatomy and vasculature of the wall of uterus simplex.

FUNCTIONAL ANATOMY OF THE MAMMALIAN FEMALE REPRODUCTIVE SYSTEM

uterine endometrium is made of uterine glands, which are shed during menstruation. The stratum basilis is not shed during menstruation but contains blood vessels that produce part of the menstrual flow.

The vascular supply of the uterus comes from the uterine artery that branches off the iliac artery. The **uterine artery** runs along the longitudinal axis of the uterus. **Arcuate arteries** branch off the uterine artery. These arteries have a circular arrangement as they invade the uterine perimetrium. The **radial arteries** branch off the arcuate arteries and penetrate deep into the myometrium. The **straight arterioles** branch off the radial arteries and terminate in the stratum basilis. These arterioles supply nutrients for the stratum basilis during endometrial regeneration. The **spiral arterioles** branch off the straight arterioles and innervate the stratum functionalis. During the menstrual cycle these arterioles constrict cutting off the blood supply to the stratum functionalis.

The microanatomy of the uterine cervix is similar to that found in the fundus and corpus. The cervical myometrium is thinner and the endometrium is not shed during menstruation. Glands lining the uterine cervix secrete mucous to varying degrees during the menstrual cycle. During specific stages of the menstrual cycle, mucous can form a plug that retards sperm movement through the cervical canal. Mucous can also prevent infectious organisms from entering the uterine cavity.

Anatomy of the Vagina

The human **vagina** is approximately 10 cm in length and lies between the urinary bladder and rectum. The length and diameter of the vagina changes during sexual arousal. Muscles of the pelvic floor can also voluntarily control the diameter of the **vaginal canal**. The vagina functions as a passageway for menstrual flow, as a receptacle for the penis during coitus and as a birth canal. A recess called the **fornix** circumscribes the external os. This recess allows support of the diaphragm contraceptive.

The microanatomy of the vagina is similar to that of other tube structures of the reproductive tract. The vagina contains an inner mucosal lining, a muscularis and an outer adventitial layer of connective tissue. The epithelial lining of the vagina consists of many layers of flattened cells. Changes in the condition of these cells during different phases of the menstrual cycle can be detected by swabbing the lining and viewing the cells microscopically. Estrogen changes the vaginal epithelium from stratified cuboidal to stratified squamous. Stratified squamous is considerably more resistant to infection and trauma.

The vagina contains a number of different types of bacteria, fungi and protozoa. These microorganisms are important in maintaining the vaginal environment. The vaginal epithelium accumulates large amounts of glycogen under the influence of progesterone. As these cells die and are sloughed into the vaginal cavity, they release their glycogen. Certain bacteria then metabolize the glycogen into lactic acid rendering the vaginal environment acidic. The acidic nature of the vagina retards yeast infection. If women take certain antibiotics that destroy these bacteria, the vagina becomes more basic often leading to a yeast infection.

THE EXTERNAL GENITALIA

Anatomy of the External Genitalia

The external genitalia of the female consist of the mons pubis, labia majora, labia minora, vaginal introitus, hymen and clitoris (Figure 4.8). These structures are collectively called the **vulva** and vary in appearance

FUNCTIONAL ANATOMY OF THE MAMMALIAN FEMALE REPRODUCTIVE SYSTEM

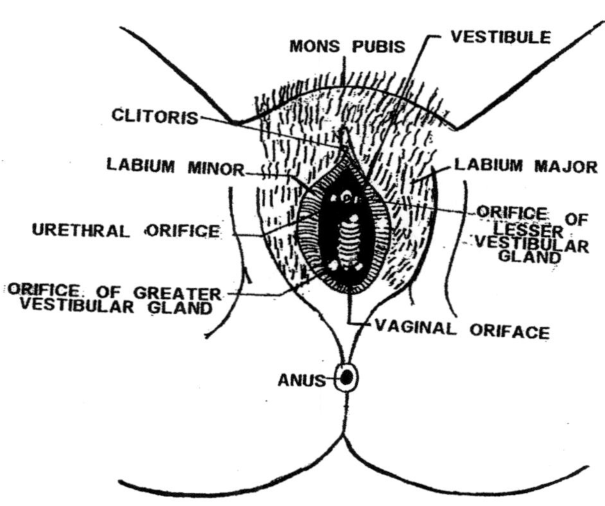

Figure 4.8. Anatomy of the external genitalia of the human female.

between species as well as between females within a species. The **mon pubis** in human females is a cushion of fatty tissue covered by skin and pubic hairs that lie over the pubic symphysis. The pubic hair forms the shape of an inverted pyramid and may extend as high as the navel. The mon pubis has many touch receptors.

The **labia majora** are fleshy folds of tissue that extend down from the mon pubis and surround the vaginal and urethral orifices. These lips contain fat and the pigmented skin has some pubic hairs, sweat and oil glands, and a few touch and pressure receptors. The labia majora are homologous to the male scrotum.

The **labia minora** are paired folds of tissue located under the labia majora. The color ranges from light pink to brownish black. These folds cover the vagina and the urethral orifices. The hairless skin of the labia minora contains oil glands but no sweat glands.

The **vestibule** is the cavity between the folds of the labia minora. Most of the vestibule is occupied by the opening of the vagina. In women that have not had coitus, the vaginal opening is often partially covered by a membrane of connective tissue known as the **hymen**. This tissue is often torn during the first coitus accompanied by minor pain and bleeding. The hymen can also be torn by a sudden fall, active participation in sports or by insertion of a tampon. In a few women with a flexible hymen, the hymen may still persist after the first coitus. In rare conditions, the hymen may completely cover the vaginal opening and may block menstrual flow at the onset of puberty. Surgery is needed to alleviate this condition.

The **urethral orifice** is the opening where urine is eliminated from the body. Below and to either side of the urethral opening are the openings of two small ducts leading to the **lesser vestibular (Skene's) glands**

FUNCTIONAL ANATOMY OF THE MAMMALIAN FEMALE REPRODUCTIVE SYSTEM

[Alexander Skene, U.S. gynecologist, 1837 – 1900]. These glands are homologous to the male prostate and secrete a small amount of fluid. At each side of the vaginal orifice are openings to the **greater vestibular (Bartholin's) glands** [Casper Bartholin, Danish anatomist, 1655 – 1738]. These glands secrete mucous and are the female's equivalent to the bulbourethral glands of the male.

The **clitoris** is a structure that lies at the upper junction of the labia minora above the urethral orifice. The average length of the clitoris in the human female is about 2.5 cm with a 5mm diameter. The clitoris is rich in pressure and temperature receptors, but has few touch receptors. The cylindrical portion of the clitoris contains a pair of corpora cavernosa that fill with blood during sexual excitement causing the clitoris to become erect. The corpus spongiosum is not present in the clitoris, but is represented in the female as the labia minora. The glans is the enlarged end of the clitoris and is partially covered by the prepuce.

The evolution of the human clitoris has fascinated reproductive physiologist for many years. The clitoris is neither large, brightly colored or selectively displayed during courtship. Clitoral design combined with its sensitivity suggests that it may be a product of female mate choice. The clitoris may distinguish male fitness that would include the physical endurance needed for long energetic sex and the mental fitness to understand the physical and psychological needs of his mate.

HUMAN BUTTOCKS AND WAIST

The Evolution of the Human Buttocks and Waist

Chimpanzees and other great apes have small, hairy, flat rumps with tough skin patches. The advent of bipedalism in hominids redesigned the muscles in the legs and buttocks. Muscles in the upper leg area became more powerful by enlargement and thus gave humans rounded buttocks. The human female deposits extra fat tissue in the hip, buttocks and upper thighs. Fat deposition in these areas is a unique feature that reveals gender difference. Fat deposition in the area of the female buttocks is age dependent. No difference exists in the amount of fat in the area of the buttocks between prepubertal boys and girls. However, at puberty human females accumulate fat in the hips, buttocks and thighs over a period of a few years. Size and protuberance of the buttocks of young women peaks around the time of greatest fertility, gradually decreasing with increasing age.

When the size of the hip and waist are measured in men, the waist to hip ratio is approximately 0.9. This is similar to measurements obtained for prepubertal girls and postmenopausal women. Young, nonpregnant, fertile females have a waist to hip ratio of approximately 0.7. When allowed to choose, men prefer women with a low waist to hip ratio. Therefore, the distribution of fat in the human female may have evolved as an indicator of youth, health and fertility in response to male mate choice.

ENDOCRINOLOGY OF THE MENSTRUAL CYCLE

Introduction

A monthly rhythmic change in the secretion of hormones and corresponding changes in the ovary and other sexual organs characterize the normal reproductive life of spontaneously ovulating mammalian females. This pattern is called the **female sexual cycle**. The **menstrual cycle** is a sexual cycle where the stratum functionalis is shed on a cyclical basis.

<u>Gonadotropic Releasing Hormone Pulse Generator of the Hypothalamus</u>

Hypothalamic neurons that secrete GnRH change their cycle of electrical activity hourly from 1000 to 3000 spikes per minute. This cyclic activity is referred to as the **GnRH pulse generator** (Figure 4.9). Biogenic amines appear to modulate the pusatile release of GnRH. Norepinephrine appears to have a stimulatory effect (an increase in pulse frequency and amplitude) on GnRH discharge whereas dopamine and serotonin have an inhibitory effect. Changes in the electrical cyclic activity of the GnRH pulse generator corresponds closely to the release of gonadotropins from the pituitary.

Changes in LH pulse frequency and amplitude occur during the human menstrual cycle. During the early follicular phase of the human menstrual cycle, the LH pulse frequency is approximately every 90 minutes with pulse amplitudes of 6.5 international units per liter (IU/L). The LH pulse frequency occurs approximately every 60 minutes to 70 minutes during the late follicular phase with a an amplitude of 7.2 IU/L. At the time of ovulation, GnRH pulse frequency occurs within a narrow range of one per hour. If the GnRH pulse frequency is less than 0.5/ hr, or greater than 1.5/ hr, no follicular maturation or ovulation occurs. During the early luteal phase of the human menstrual cycle, the LH pulse frequency occurs at approximately every 100 minutes with amplitude of 15.0 IU/L. The LH pulse frequency changes to one every 200 minutes during the late luteal phase with amplitudes of 8.0 IU/L.

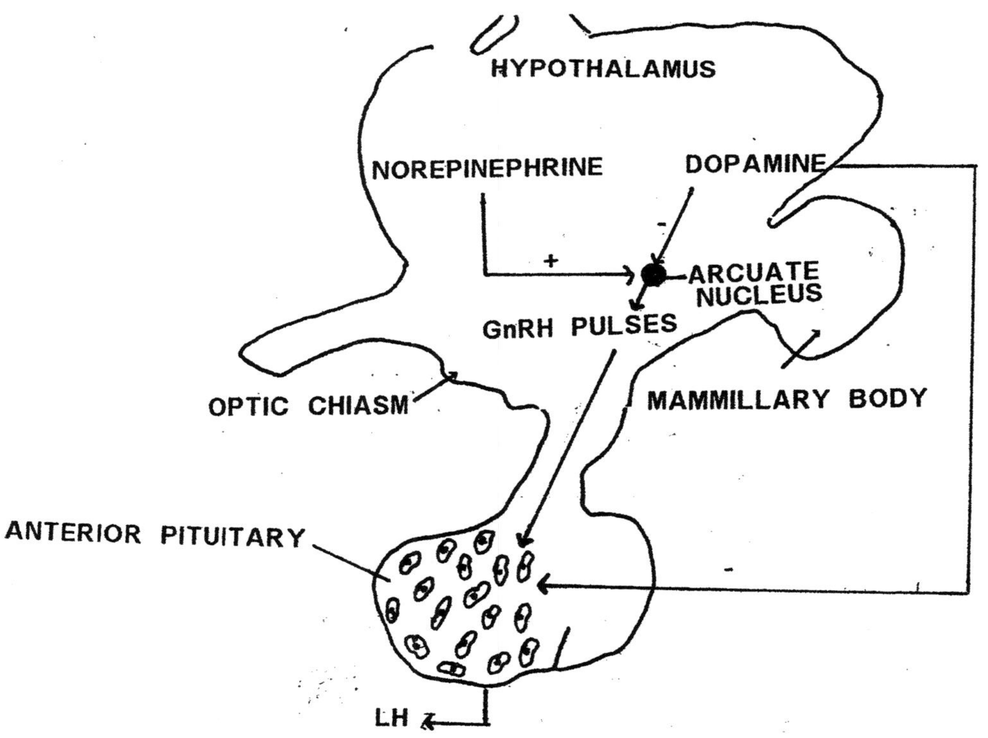

Figure 4.9. The GnRH pulse generator. Changes in the release of GnRH and LH.

FUNCTIONAL ANATOMY OF THE MAMMALIAN FEMALE REPRODUCTIVE SYSTEM

<u>Phases of the Menstrual Cycle</u>

The menstrual cycle can be divided into three phases called the menstrual phase, follicular phase, and the luteal phase (Figure 4.10). The average length of the menstrual cycle in humans is 28 days with a range of 25 days to 35 days. In western populations, 0.5% of human females have menstrual cycles that average less than 21 days in length, whereas 0.9% of human females have menstrual cycles that average greater than 35 days in length. The length of the follicular phase varies between women and often varies from cycle to cycle within a given individual. This accounts for most of the variation in the length of the menstrual cycle. The length of the luteal phase is about 14 days no matter what the length of the other phases of the cycle. Therefore, ovulation occurs approximately 14 days before the onset of menstruation.

The **menstrual phase** of the human menstrual cycle is generally considered to be day 1 to day 4.5 of the cycle (Figure 4.10). Plasma FSH and LH levels are low, but start to steadily increase. Follicle stimulating hormone is responsible for recruitment and development of follicles for the next cycle. The corpus luteum of the previous cycle has regressed causing a significant decrease in plasma concentrations of estrogen and progesterone. This decrease in plasma steroid concentrations is responsible for constricting uterine endometrial blood vessels that decrease the amount of oxygen and nutrients supplying the stratum functionalis. Starvation of the stratum functionalis results in menses of this layer of the endometrial lining.

The **follicular phase** of the human menstrual cycle is day 5 to day 14 (Figure 4.10.). Follicle stimulating hormone is responsible for development of recruited follicles in which some eventually become atretic. During the early follicular phase, activin produced by granulosa cells of immature follicles can enhance the action of FSH (Figure 4.11). Activin increases the aromatase activity of follicular granulosa cells. Activin also enhances FSH and LH receptor formation. Activin decreases androgen production by theca cells. The uterine endometrium proliferates and uterine glands enlarge due to the influence of estrogens. The endometrium becomes richly supplied by blood vessels. Estrogen also increases uterine muscular activity.

In the late follicular phase, increased production of inhibin-B and decreased synthesis of activin by granulosa cells promotes androgen production by theca cells in response to LH stimulation (Figure 4.12). Activin prevents luteinization of follicular cells prior to ovulation. Granulosa cells of the dominant follicle acquire the highest aromatase activity producing the highest concentration of estradiol 17-*B*. Plasma levels of estradiol 17-*B* peak in the blood approximately 36 hours to 24 hours prior to ovulation. A surge in plasma LH concentration occurs approximately 10 hours to 12 hours prior to ovulation. Approximately 66% of women have the peak plasma LH surge between 3:00 AM to 8:00 AM in the morning. The surge in plasma LH concentration is due to the positive feedback of estradiol 17-*B* on the hypothalamus and the anterior pituitary. Shortly after the surge in plasma LH concentration, levels of estradiol 17-*B* decrease in the blood. The LH surge initiates meiosis of the primary oocyte resulting in the production of a secondary oocyte arrested in metaphase II. Expansion and luteinization of cumulus granulosa cells is initiated following the plasma surge of LH. Increased synthesis of prostaglandins and progesterone also occurs in response to the plasma surge of LH concentration. Prior to ovulation, the richly vascularized dominant follicle develops a hyperemic appearance and looks like a blister on the ovarian surface. Granulosa cells acquiring lipid inclusions and theca cells become vacuolated.

The **luteal phase** of the human menstrual cycle is considered to be day 14 to day 28 (Figure 4.10). The

FUNCTIONAL ANATOMY OF THE MAMMALIAN FEMALE REPRODUCTIVE SYSTEM

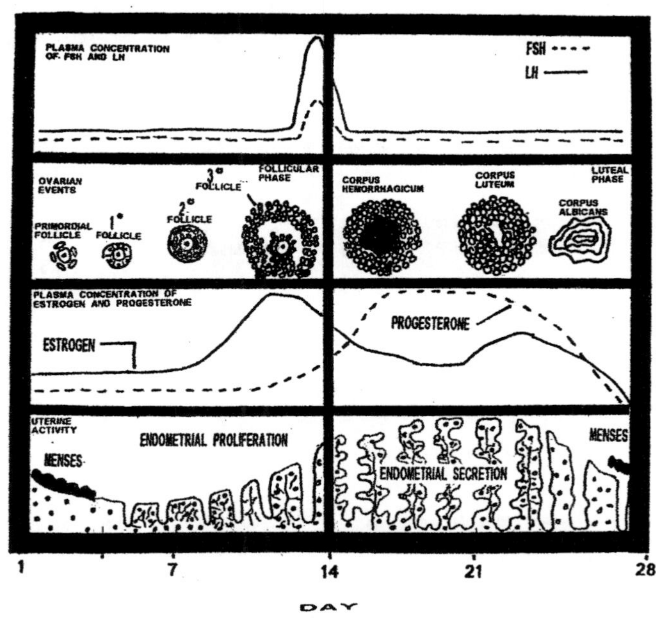

Figure 4.10. The human menstrual cycle.

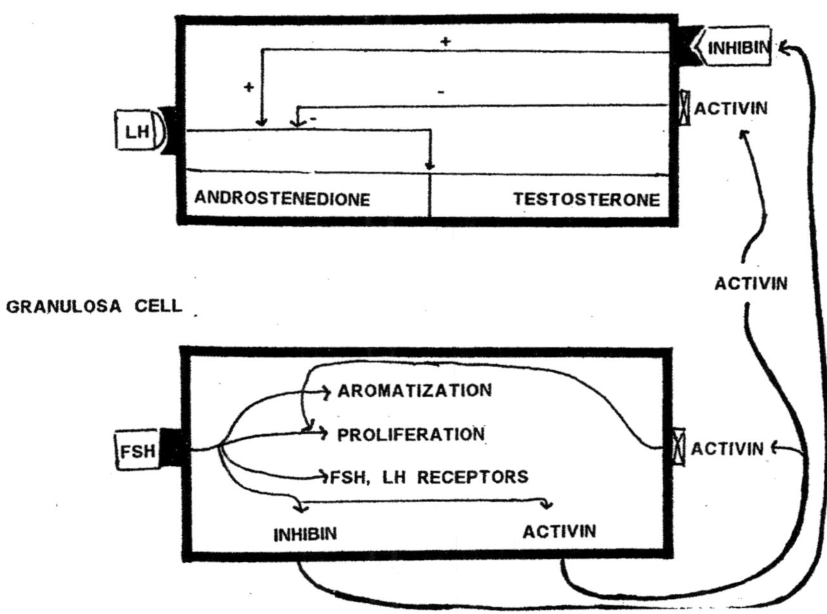

Figure 4.11. Ovarian cellular events during the early follicular phase.

FUNCTIONAL ANATOMY OF THE MAMMALIAN FEMALE REPRODUCTIVE SYSTEM

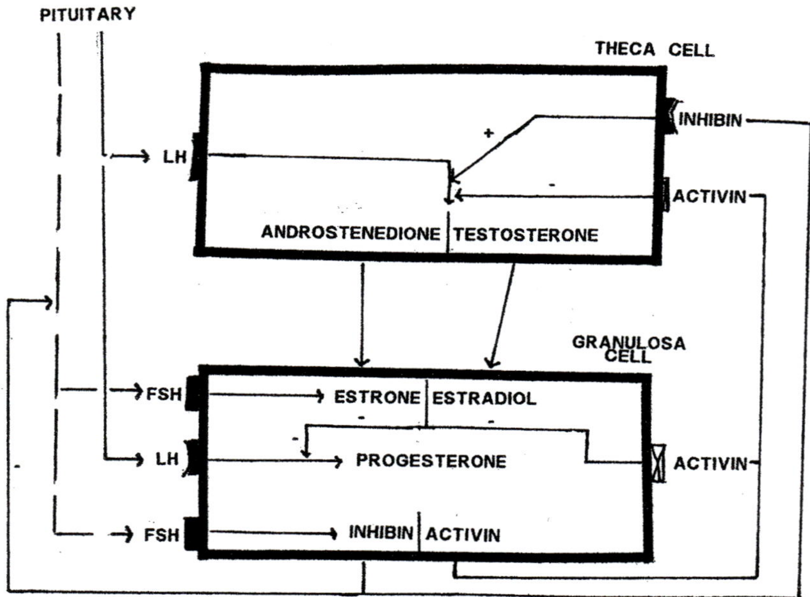

Figure 4.12. Ovarian cellular events during the late follicular phase.

average length of the luteal phase is approximately 14 days with a range of 11days to 17 days. In western society, 5% to 6% of human females have short luteal phases possibly due to active luteal mechanisms. After ovulation, granulosa cells in the remnant follicle continue to enlarge for 3 days. Vascularization of the granulosa cells that was initiated after the plasma LH surge, peaks approximately 8 days to 10 days post ovulation giving rise to the corpus luteum. Angiogenesis of the corpus luteum occurs in response to LH mediated growth factors such as epidermal growth factor. Normal luteal function requires optimal preovulatory follicular development. The degree of preovulatory follicular develop determines the extent of luteinization and the functional capacity of the corpus luteum.

The corpus luteum produces moderate levels of estrogen and elevated levels of progesterone. The secretion of estrogen and progesterone is episodic and correlates with LH secretory pulses. Plasma progesterone concentration rises sharply, peaking approximately 8 days after ovulation. Estrogen and progesterone have a negative feedback effect on the secretion of FSH and LH thus suppressing follicular growth. Progesterone can locally suppress follicular growth. Progesterone is also responsible for stimulating endometrial glandular secretion. Progesterone is also responsible for the quiescence of the uterine musculature. Inhibin A plasma concentration peaks during the mid-luteal phase and is responsible for suppressing FSH levels to nadir levels.

If pregnancy does not occur, the corpus luteum regresses resulting in a decrease in the plasma levels of estrogen and progesterone. This results in constriction of blood vessels of the uterine endometrium resulting in menses. The cause for corpus luteal regression varies between species. Human luteal regression may possibly be due to the deactivation of the G protein adenylate cyclase system that occurs in response to LH stimulation. During luteolysis there is a direct increase in the expression of genes involved in the production of metalloproteinases (MMPs). Stimulation of the corpus luteum by hCG decreases the expression of MMPs genes resulting in survival of the corpus luteum.

FUNCTIONAL ANATOMY OF THE MAMMALIAN FEMALE REPRODUCTIVE SYSTEM

A progressive and rapid increase in GnRH pulsatility occurs at the luteal/follicular transition period of the human menstrual cycle. The pulsatility of LH and FSH increase 4.5X and 3.5X, respectively. The increase in the mean concentration of LH and FSH secreted increases approximately 2X and 3.5X, respectively. Plasma inhibin B concentration begins to increase shortly after the increase in plasma FSH concentration peaking 4 days after the maximal concentration of plasma FSH. Maximal plasma activin concentration occurs at menses.

Evolution of Menstruation

If pregnancy does not occur during a given sexual cycle, the uterine endometrium can undergo one of three alternatives: (1) The endometrium can be maintained for the next sexual cycle and a possible pregnancy. (2) The endometrium can be completely absorbed (3) The endometrium can be sloughed off (menstruation) to the external environment. Old World primates, including humans, have chosen menstruation as the way of dealing with the endometrium when pregnancy does not occur.

The evolution of uterine menstruation has fascinated reproductive physiologists for many years. Although the reason for menstruation is currently unclear, several hypotheses have been proposed. Menstruation may possibly occur to cleanse the female reproductive tract of potential pathogens that may have entered during copulation. Retaining the nutrient-rich endometrium would provide a prime habitat for bacterial infection. Shedding of the endometrium would eliminate this possibility thus preventing infections of the upper reproductive tract and abdomen. Menstruation may possibly occur because maintaining or absorbing the uterine endometrium is a metabolically costly process. During the luteal phase of the menstrual cycle, the endometrium of the human uterus is thicker and better developed compared to that of many other mammalian species. To maintain the endometrium for the possibility of pregnancy during the next sexual cycle would involve a high metabolic cost. Therefore, it may be metabolically cheaper for humans to expel the endometrium to the external environment and rebuild a new uterine endometrium for the possibility of pregnancy during the next sexual cycle.

THE ESTROUS CYCLE

Overview

Domestic animals such as sheep, cattle and swine exhibit a female sexual cycle referred to as the **estrual cycle**. The hormonal profiles of the estrual cycle are very similar to that observed for the menstrual cycle. However, there is no shedding of the uterine endometrium during the estrual cycle.

Timing of the estrous cycle differs from that observed for the menstrual cycle. In the estrual cycle ovulation occurs at day 0 compared to day 14 for the menstrual cycle.

The estrous cycle of most domestic animals is between 16 days to 21 days in length (Table 4.1). The length of estrous or the cyclical period of sexual activity can vary from 18 hours to 19 hours in cattle, to 4 to 8 days in the mare (Table 4.1). In most domestic animals the time of ovulation occurs near the end of estrous (Table 4.1).

FUNCTIONAL ANATOMY OF THE MAMMALIAN FEMALE REPRODUCTIVE SYSTEM

ANIMAL	LENGTH OF ESTROUS CYCLE (DAYS)	DURATION OF ESTROUS (HOURS)	TIME OF OVULATION (HOURS FROM ONSET OF ESTROUS)
SHEEP	16 - 17	24 - 36	24 - 30
GOAT	21	32 - 40	30 - 36
PIG	19 - 20	48 - 72	35 - 45
COW	21 - 22	18 - 19	29 - 30
MARE	19 - 25	96 - 192	1 - 2 DAYS BEFORE THE END OF ESTROUS

Table 4.1. The estrous cycle. Estrous and ovulation of major domestic mammalian species.

MAMMALIAN FEMALE REPRODUCTIVE CHARACTERISTICS

Spontaneous Ovulation

Mammalian species that are **spontaneous ovulators** have a defined ovarian cycle. **Polyestrous nonseasonal breeders** cycle around the year at regular intervals (e.g. cattle and pigs). **Polyestrous seasonal breeders** cycle only during the breeding season (e.g. horse, sheep and goat). Horses are long-day breeders being sexually receptive from spring to fall. Sheep and goats are short-day breeders being sexually receptive from late summer to early winter. Monoestrous seasonal breeding mammals (e.g. dog) shows one sexually receptive cycle during the breeding season. Although the dog has often been considered a monoestrous seasonal breeder, variation in dog breeds and environmental influences often have dogs showing several receptive cycles during the year.

Several spontaneous ovulating mammalian species (e.g. cattle, swine, sheep and horses) have short follicular phases. These species show significant follicular antral expansion during the previous luteal phase. This is possible because FSH and LH do not fall to nadir levels during the luteal phase. Complete antral expansion, ovulation and a complete luteal phase occupy more than one complete estrous cycle.

Rats and mice can show abbreviate follicular and luteal phases. The ovarian cycles of these rodent species are condensed compared to the large domestic animal species described above. The cycle length of

FUNCTIONAL ANATOMY OF THE MAMMALIAN FEMALE REPRODUCTIVE SYSTEM

mice and rats depend on mating of the female. If the female has an infertile mating at ovulation, the luteal phase is approximately 11 days to 12 days in length (similar to that observed in swine). If the female fails to mate at ovulation, the luteal phase will be approximately 2 days to 3 days in length. The poorly developed corpus luteum produces progesterone in small amounts. The sexual cycle of rodents and mice can be explained as follows. Penile stimulation of the cervix generates sensory bioelectrical information that is transmitted to the central nervous system. The central nervous system stimulates the release of prolactin from the anterior pituitary. Prolactin is part of the luteotropic complex involved in the maintenance of the corpus luteum. Without the release of prolactin from the anterior pituitary, the life of the corpus luteum is shortened to approximately 2 days to 3 days. The evolutionary advantage of ovarian cycles with abbreviated follicular and luteal phase is an increase in reproductive efficiency.

<u>Induced Ovulation</u>

Mammalian species that are **induced ovulators** generally have a removed luteal phase in which ovulation is induced by penile or mechanical stimulation of the cervix. Alpacas and llamas are induced ovulators that are continuously receptive during the breeding season until mating occurs. Cats are seasonal breeding induced ovulators with reoccurring periods of receptivity. Feral cats often have breeding seasons in the spring and fall. Cats maintained in environments with controlled light and temperature tend to show a decreased sexual receptivity from October to December.

Rabbits are induced ovulators. Histological examination of the rabbit ovary shows waves of expanded follicles. Plasma concentration of estrogen is elevated, whereas the plasma concentration of progesterone is extremely low. If the doe is mated to a vasectomized buck (or the cervix is stimulated manually), she will ovulate approximately 10 hours to 12 hours later and has a luteal phase (pseudopregnancy) of approximately 12 days in length. If the vasectomized buck is left with the doe, she will have a sexual cycle of approximately 14 days with a follicular/luteal pattern of 2 days and 12 days, respectively.

THE FEMALE SEXUAL RESPONSE CYCLE

<u>General</u>

When exposed to the appropriate thoughts and stimuli, the female will elicit a **sexual response cycle**. The sexual response cycle of the female is divided into an excitement phase, plateau phase, orgasmic phase and a resolution phase. Below the sexual response cycle for the human female is discussed.

<u>Excitement Phase</u>

The **excitement phase** is usually initiated within 30 seconds after exposure to erotic stimuli. Physiological changes are noted in the reproductive tract, external genitalia, breast and skin. The inner two thirds of the vaginal barrel begins to increase in length and width. The vaginal walls become engorged with blood taking on a dark coloration. Vaginal lubrication starts to occur. The body of the uterus ascends (the tenting effect) pulling the cervix away from the vagina thus increasing vaginal length. The size of the uterus starts to increase in response to vasocongestion. Rapid irregular uterine contractions occur that are not painful. The labia minora engorge with blood and their size increases. The labia majora flatten out and retract from the midline. The nipples become erect and the breast increases in size. The sexual flush occurs in 75% of women due to dilation of peripheral blood vessels. An increase in muscular tension can be noted.

FUNCTIONAL ANATOMY OF THE MAMMALIAN FEMALE REPRODUCTIVE SYSTEM

Plateau Phase

During the **plateau phase** sexual excitement continues and becomes enhanced. The outer one third of the vaginal wall becomes greatly engorged with blood causing this portion of the vaginal cavity to become reduced in size. The clitoris retracts to become completely covered by the clitoral prepuce. The length of the clitoris decreases by approximately 50%. Clitoral stimulation at this point can occur by stimulation through the prepuce as well as by pressure applied to the labia minora. Uterine fibrillations continue and increase in intensity. Tenting of the uterus becomes more pronounced. The nipples continue to erect and the areolas become darker in coloration. Blood pressure, cardiac rate and depth of pulmonary ventilation increase considerably. Tension of skeletal and smooth muscles becomes more pronounced.

Orgasmic Phase

Orgasm comes from the word orgasmos, meaning "to swell or be lustful". Strong muscular contractions of the outer third of the vaginal barrel occur during the **orgasmic phase**. The first contractions last approximately 2 to 4 seconds and are followed by rhythmic contractions approximately 0.8 seconds apart. The intensity of the initial contraction is stronger than the later ones. The rectal sphincter exhibits rhythmic contractions that are also 0.8 seconds apart.   The inner two thirds of the vaginal barrel expands to facilitate the movement of the penis. Rhythmic contractions of the uterus occur probably brought on by the release of oxytocin from the posterior pituitary. The sex flush peaks in intensity and distribution. The cardiac rate, blood pressure as well as the rate and depth of pulmonary ventilation is similar to that observed in the male. There is a great release in neuromuscular tension and strong involuntary muscular contractions. Clawing motions of the hands and feet can sometimes be observed.

The major difference between males and females in the sexual response cycle is that females do not have a refractory period immediately after orgasm. Some women fluctuate from the orgasmic phase to the plateau phase, then back to the orgasmic phase. Many women report that the later orgasmic phases are more intense than the initial one. A few women can have "status orgasmus" which is sustained orgasm.

The experience of orgasm can vary within women and between women. Women experience a sensation of "suspension" lasting only a brief moment followed by a period of intense sexual awareness oriented at the clitoris and radiating up into the pelvis. There is also a sensation of warmth that begins in the pelvis and spreads to other parts of the body. Pelvic throbbing can also be felt.

There are several kinds of orgasms. Clitoral orgasm is caused by stimulation of the clitoris. Vaginal orgasm is believed to be due to direct stimulation of the vaginal wall. The **Graefenberg spot** [Ernst Graefenberg, German gynecologist in the US, 1881 – 1957] is a small region in front of the vagina. Stimulation of this area can produce sexual arousal and intense contractions of the uterus and the pubococcygeus muscle.

Resolution Phase

In the **resolution phase**, the female physiological functions return to normal values. Vaginal contractions cease and the clitoris leaves its contracted position. Cardiac rate, blood pressure as well as the rate and depth of pulmonary ventilation start to return to normal values. The labia minora begin to take on a pink coloration in response to decreased vasocongestion. The internal cervical os dilates immediately after orgasm to allow the movement of spermatozoa into the uterus. Within 5 to 10 minutes post orgasm, muscular tension decreases. The breasts decrease in size and the uterus returns to its normal size and position. The labia majora returns to its original state within one hour.

GAMETE TRANSPORT AND FERTILIZATION

TRANSPORT OF SPERMATOZOA IN THE FEMALE REPRODUCTIVE TRACT

Site of Semen Deposition

The site of semen deposition varies between species. The semen of rodents, pigs and horses is deposited directly into the uterus. In rabbits and primates, semen is deposited in the vagina close to the external os of the cervix.

Spermatozoa become trapped in the coagulum of the seminal plasma in vaginally deposited semen. In order to ascend the female reproductive tract, the semen must liquefy and the vaginal wall must contract to push the spermatozoa in the direction of the cervix. Spermatozoa have been shown to be in the oviduct within 5 minutes post ejaculation. This suggests that some spermatozoa are propelled directly into the cervix shortly after the ejaculatory process.

Human spermatozoa thrive at a seminal pH between 7.2 to 7.8. However, the pH of human vaginal secretions is around 5.7. Spermatozoa are also attacked by leukocytes found in vaginal secretions. This shows that the vagina is a hostile environment for spermatozoa.

Transport of Spermatozoa through the Cervix

The anatomy of the cervix and the composition of the cervical mucous is important in the transport of vaginally deposited spermatozoa.   These factors are discussed below.

The cervical canal follows an irregular course with numerous deep crypts in the wall (Figure 5.1). The mucous membrane produces copious amounts of mucous. The mucous is composed of mucin, inorganic salts, glucose, maltose, mannose, amino acids, peptides, proteins and lipids. The mucin is a peptide with a fibrillar system of carbohydrate molecules linked directly to the disulfide bridges or indirectly to the peptide.

Cervical mucous becomes thin and watery near the time of ovulation. It can often be seen dripping from the vulva of domesticated farm animals. Watery cervical mucous is caused by the macromolecular fibrils becoming oriented in parallel chains or micelles. This allows the spermatozoa to swim through the mucous. Under the influence of progesterone, the macromolecular fibrils and micelles loose their parallel arrangement. This causes the mucous to become viscous and rubbery which is not conducive to spermatozoa transport.

After vaginal semen deposition, most spermatozoa enter the crypts of the cervical glands and remain there for a considerable period of time (72 hours for humans). The cervical glands provide a favorable environment and protect the spermatozoa from phagocytic attack. The glands also act as a reservoir by releasing spermatozoa over a period of time.

Both sperm motility and cervical contraction are required to transport spermatozoa through the cervix. Sperm can traverse the cervical mucous at a rate of 0.1 mm/min to 0.3 mm/min. The number of

GAMETE TRANSPORT AND FERTILIZATION

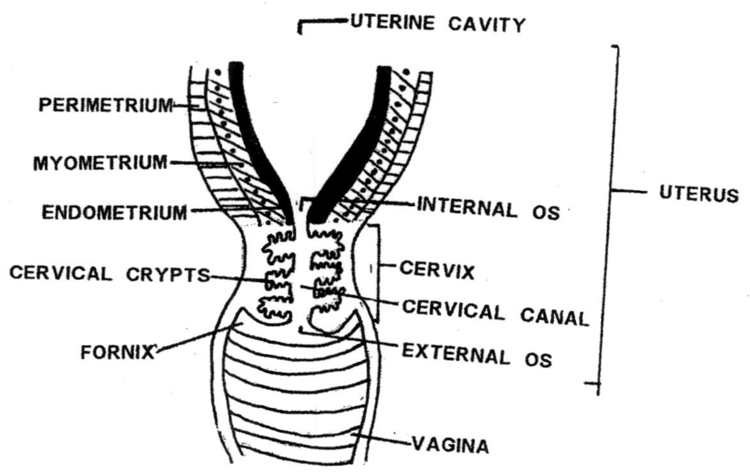

Figure 5.1. Anatomy of the cervical region of uterus simplex.

spermatozoa reaching the upper regions of the cervix increases over time, but is small compared to the number of spermatozoa ejaculated.

Transport of Spermatozoa through the Uterus

The transport of spermatozoa through the uterus in most mammalian species takes 15 minutes to 45 minutes. This rapid transport occurs in response to vigorous uterine contractions. In some species (e.g. sheep) uterine motility appears to be stimulated by the presence of prostaglandins in the semen. In other species (e.g. humans and cattle) oxytocin released from the posterior pituitary during coitus appears to be primarily responsible for contraction of the uterine myometrium.

For most mammalian species, the lifespan for spermatozoa in the uterine cavity is approximately 24 hours. Most of the sperm cells are expelled back into the cervix just a few hours after entering the uterine cavity.

Many spermatozoa succumb to phagocytosis by leukocytes. Spermatozoa that manage to find their way into uterine glands are protected from phagocytosis and other lethal agents. In a few mammalian species, spermatozoa can maintain their fertility for a considerable period of time. In bats, spermatozoa can maintain their motility for several months. In mice, two successive pregnancies can occur from one ejaculate.

Transport of Spermatozoa through the Oviducts

In many mammalian species, waves of uterine contractions can transport spermatozoa into the oviducts within several minutes post insemination. These spermatozoa are usually damaged because of the shearing forces created in response to uterine contraction. The oviduct contains secretory cells and their secretions fill the lumen of the oviduct. Luminal flow is primarily towards the ovary and the peritoneal cavity (Figure 5.2). The oviductal fluid confined to a shallow layer over the epithelial surface, flows in the direction of the uterus. When this surface fluid reaches the ampullary–isthmus junction, the fluid reverses direction and the flows back as an axial stream towards the open end of the oviduct and into the peritoneal cavity.

In order for spermatozoa to enter the oviduct, they must pass through the uterotubal junction. During most of the female cycle, the uterotubal junction is closed and will not permit the passage of spermatozoa. However, shortly after ovulation the uterotubal junction opens and spermatozoa can pass into the oviduct.

The isthmus of the oviduct of cattle, humans, mice, hamsters and pigs has a luminal diameter approximately the size of a sperm head and is filled with mucous. The inner mucosal lining of the isthmus oviduct is arranged in branching folds thus creating intricate passages and blind alleys. The transport of spermatozoa through the oviducts of many mammalian species appears to be a discontinuous process. Prior to ovulation, the plasma membrane of spermatozoa make contact with oviductal epithelial cells. In cattle and hamsters, capacitation and hyperactive movement of sperm cells occurs in the isthmus of the oviduct. In humans, spermatozoa continue to move through the isthmus of the oviduct. The isthmus of the human oviduct does not appear to serve as a sperm reservoir. Movement of spermatozoa into the ampullary-isthmus junction occurs by a combination of sperm motility and muscular contraction of the oviduct. The transport of spermatozoa from the ampulla into the peritoneal cavity occurs in response to their own motility, muscular contraction of the oviduct and the luminal flow of fluid from the oviduct into the peritoneal cavity.

In humans, motile spermatozoa have been found in the oviduct for as long as 85 hours post coitus. However, the fertilizing capability of spermatozoa is usually lost within 24 hours to 48 hours. Spermatozoa can remain alive in the peritoneum for at least 24 hours. Fertilization and early embryonic development have been initiated in the dog, rabbit, cow and quinea pig following intraperitoneal injection of a sperm suspension.

TRANSPORT OF THE OVUM THROUGH THE FEMALE REPRODUCTIVE TRACT

<u>Transport of the Ovum through the Oviduct</u>

The mechanism by which an ovulated ovum is transported into the oviduct is not entirely known. Movement of fimbrial cilia creates a flow of peritoneal fluid into the ostium of the oviduct. This fluid flow appears to be responsible for the transport of an ovulated ovum from the ovarian surface into the oviduct.

Transport of an ovum through the ampulla of the oviduct is dependent on the mechanical action of cilia on the corona radiata. If the cilia of the ampulla are paralyzed or the cumulus mass is removed from the ovum, ovum transport is drastically impaired. Once an ovum has reached the site of fertilization, much of the corona radiata is lost due to mechanical action of the cilia, muscular contraction of the oviduct and hyaluronidase released from spermatozoa.

Transport of an ovum through the isthmus of the oviduct occurs primarily in response to muscular contraction. Ciliary movement and fluid forces have little influence on ovum transport in this area. Muscular contraction within the isthmus is not peristaltic, but involves contraction of the circular smooth

GAMETE TRANSPORT AND FERTILIZATION

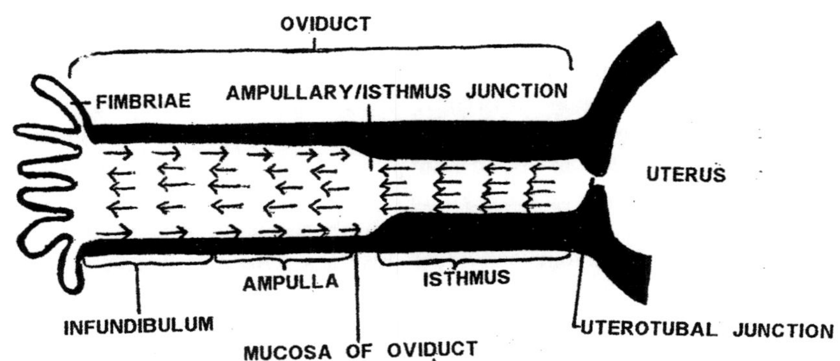

Figure 5.2. Flow of oviductal secretions near the time of fertilization.

muscle layer. This imparts a "to and fro" movement to the ovum in the net direction of the uterus. In many litter-bearing species (e.g. swine), the eggs are usually clumped a little distance from the uterus prior to entrance. This suggests that the final passage of ova into the uterus is rapid and depends on muscular relaxation of the oviduct.

The overall transport time of an ovum down through the oviduct is approximately 3 days. In swine, ovum transport time is slightly faster than 3 days, whereas in carnivore's ovum transport time is slightly longer than 3 days. In equine, an unfertilized ovum may remain in the oviduct for weeks or months, and be bypassed by subsequent ovulated or fertilized eggs.

FERTILIZATION

Capacitation of Spermatozoa

Capacitation is a process whereby the glycoproteins of the spermatozoon membrane are modified and the seminal proteins are removed. Capacitation of spermatozoa is required for fertilization of an ovum. For spermatozoa to become capacitated, they must be exposed to the female reproductive tract for several hours during the late follicular phase of the reproductive cycle. Spermatozoa in the female reproductive tract during the luteal phase do not become fertile. This suggests there are specific and distinct factors that effect sperm capacitation. If rabbit spermatozoa are exposed to the oviduct of the rat and then exposed to rabbit ova, fertilization occurs. This shows that capacitation is not a species-specific process.

Spermatozoa that are capacitated have changes in their membrane lipids, acquire the ability to undergo the acrosome reaction and become hypermotile. The precise cellular changes that take place during capacitation have been difficult to pinpoint. During capacitation of a spermatozoon, cholesterol and other sterols are removed from the plasma membrane. Phospholipid content of the spermatozoon plasma membrane is changed (Figure 5.3). Increased membrane permeability to ionic bicarbonate and ionic calcium allow these ions to influx into the spermatozoon, whereas ionic potassium effluxes from the cell. An increase in intracellular pH (because of the influx of ionic bicarbonate into the cell) and intracellular ionic calcium enhances the synthesis of c-AMP. Cyclic adenosine monophosphate activates numerous molecules of protein kinase A. Since protein kinase A cannot phosphorylate proteins on tyrosine residues, indirect method(s) must be employed. Protein kinase A stimulates protein tyrosine kinase while inhibiting phosphotyrosine phosphatase. Protein kinase A stimulates the phosphorylation of proteins on serine and threonine residues that then phosphorylate proteins on tyrosine residues. Phosphorylating proteins on tyrosine residues stimulates hypermotility of the spermatozoon and contributes to capacitation. The efflux of ionic potassium from the spermatozoon changes the membrane potential from -30 mV to -50 mV. This hyperpolarization of the spermatozoon is also important in capacitation.

<u>The Acrosome Reaction</u>

The greatest morphological transformation that capacitated spermatozoa can undergo is the **acrosome reaction** (Figure 5.4). This is a process that spermatozoa must complete in order to fertilize the egg. In most mammalian species, the acrosome reaction occurs at the zona pellucida of the ovum after the spermatozoa have penetrated the corona radiata. The spermatozoon attaches to the ZP3 receptor. This initiates the influx of ionic calcium into the spermatozoon. An increase in intracellular ionic calcium is responsible for increasing intracellular secondary messengers. An increase in intracellular pH also occurs. Fibrous (F) actin depolymerizes forming soluble monomeric actin. This allows the spermatozoon plasma membrane and outer acrosomal membrane to come into close proximity. Activation of phospholipase A$_2$ promotes fusion and vesiculation of the plasma membrane and the outer acrosomal membrane of the spermatozoon.

Acrosin and **hyaluronidase** are two important acrosomal enzymes needed during the fertilization process. Hyaluronidase breaks down the hyaluronic acid that binds together the cellular investment surrounding the ovum. The release mechanism of hyaluronidase from sperm cells is not completely understood. Progesterone stimulates the influx of ionic calcium into the spermatozoon thus causing an increase in secondary intracellular messengers. Follicular production of progesterone rises immediately prior to ovulation and it may be possible that the cumulus mass surrounding the ovum contains progesterone. Arrival of the spermatozoon at the cumulus mass may make the acrosome "leaky" allowing hyaluronidase to coat the outer surface of the spermatozoon. This would facilitate the passage of the spermatozoon through the cumulus mass surrounding the ovum. Acrosin is stored within the acrosome in a zymogenic form and on the surface of capacitated sperm. This enzyme is important in binding and penetration of the zona pellucida.

<u>Gamete Fusion and Incorporation of the Spermatozoon into the Ovum</u>

After penetration of the zona pellucida, the spermatozoon may swim around the perivitelline space for a few brief moments. Hypothetically, cell adhesion of mammalian gametes occurs in three phases (Figure 5.5). In the **initial attachment phase of mammalian gametes,** ligand 1 on the anterior inner acrosomal membrane of the spermatozoon attaches to a receptor on the oolemma. The spermatozoon then rotates

GAMETE TRANSPORT AND FERTILIZATION

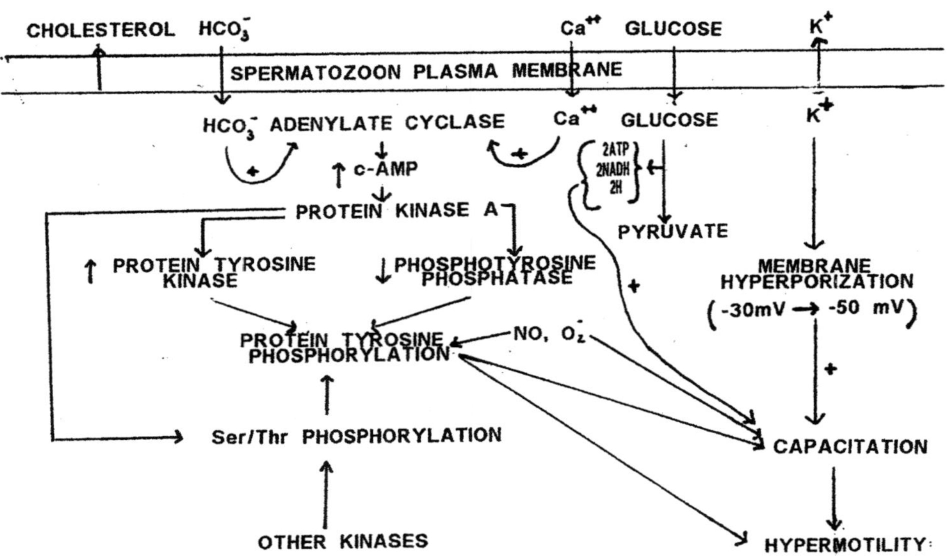

Figure 5.3. Cellular mechanism involved in the capacitation of the spermatozoon.

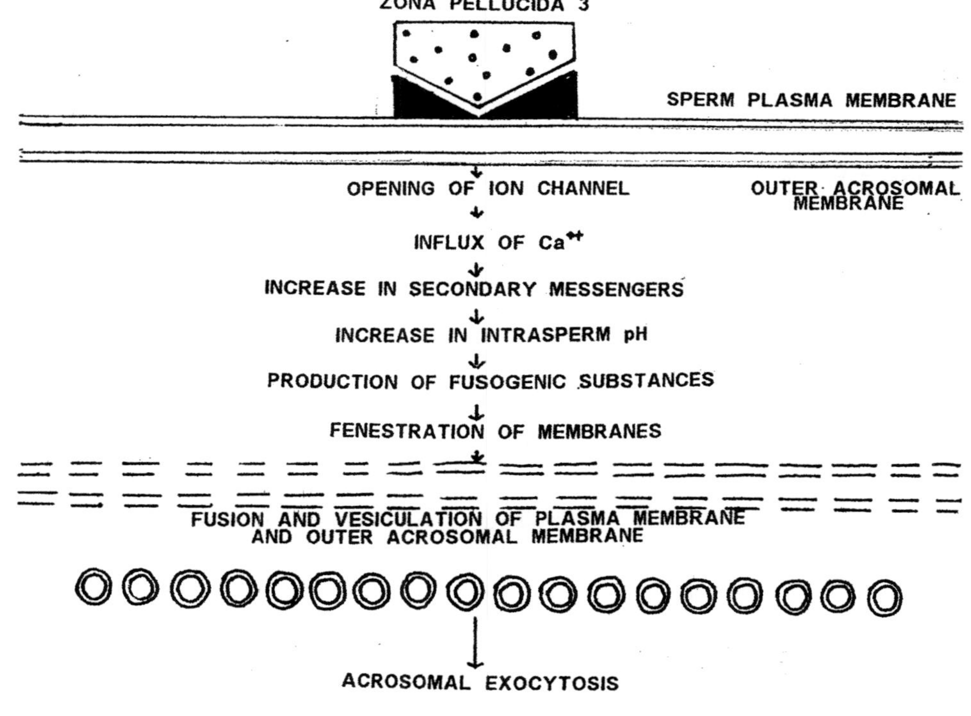

Figure 5.4. The acrosome reaction of a mammalian spermatozoon.

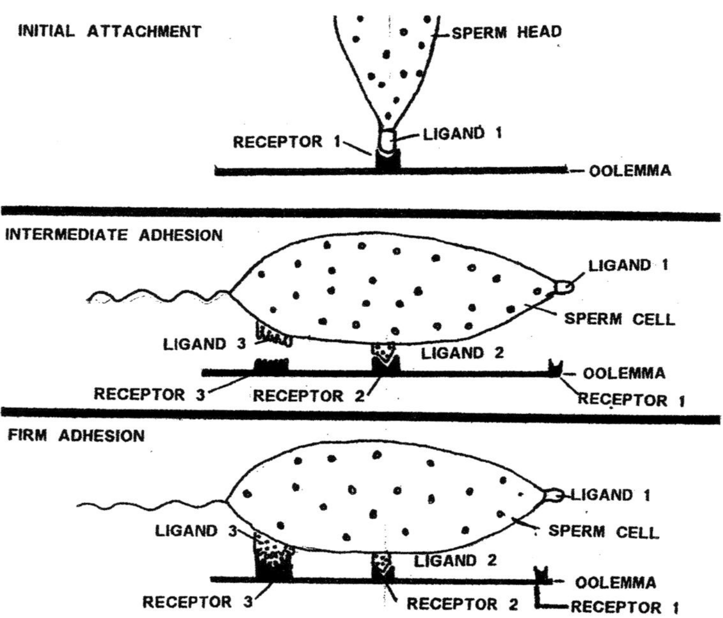

Figure 5.5. Adhesion of the mammalian spermatozoon to the oolemma.

down to the surface of the oolemma allowing **intermediate adhesion of the gametes** between ligand 2 on the equatorial region of the spermatozoon and the receptor on the oolemma. **Firm adhesion of the gametes** is accomplished by ligand 3 on the posterior region of the spermatozoon attaching to another receptor on the oolemma.

The hypothetical model for the membrane fusion between mammalian gametes is believed to occur in four steps (Figure 5.6). After firm adhesion of the spermatozoon to the oolemma, a fusion protein in the sperm cell membrane undergoes a conformational change (step 1). The conformational change in the fusion protein allows it to be inserted into the oolemma (step 2). A second conformational change of the fusion protein allows bending the membranes of the spermatozoon and the ovum (step 3). Lastly, a mixture of the bilayer components of the sperm cell membrane and the ovum results in hemifusion and formation of fusion pore (step 4). The nucleus of the spermatozoon together with variable portions of the midpiece and tail pass into the ooplasm.

<u>Activation of the Egg</u>

Interaction of the plasma membranes of the sperm and the ovum leads to a series of events collectively known as **egg activation.** Egg activation includes the initiation of oscillations in the intracellular ionic calcium concentration, continuation of meiosis of the secondary oocyte, entrance into first embryonic mitotic division and blockage to polyspermy.

According to the **sperm factor hypothesis,** activation of the egg is initiated by an unknown factor from the interior of the sperm that migrates to the ooplasm of the egg after membrane fusion. This unknown factor

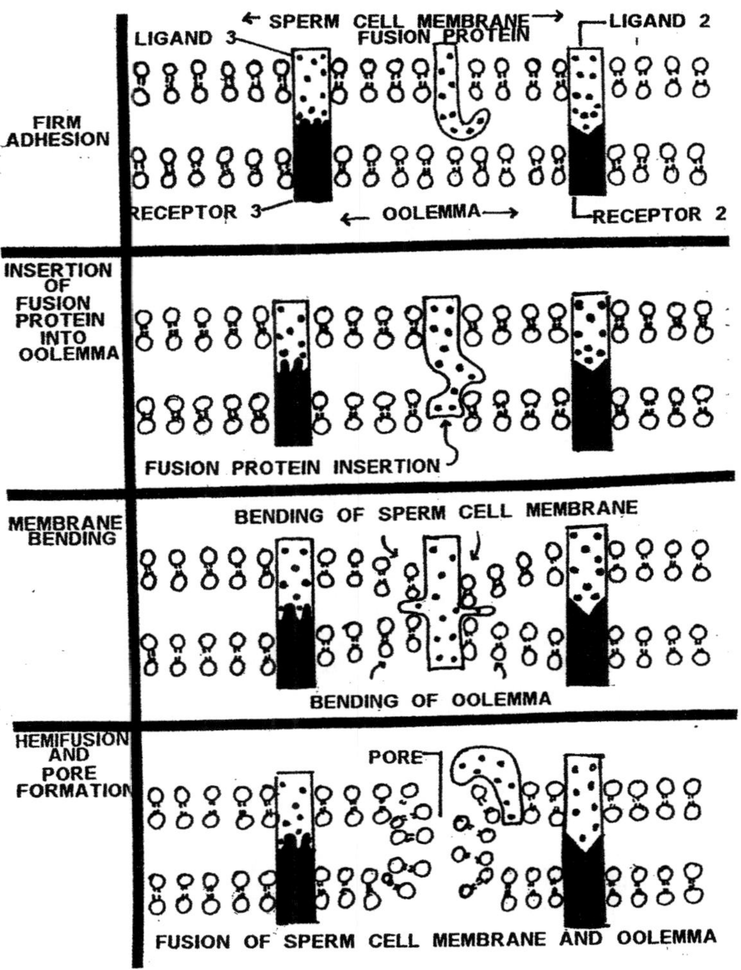

Figure 5.6. A hypothetical model for the fusion of the mammalian spermatozoon to the egg.

initiates the release of ionic calcium from intracellular stores. Increased intracellular ionic calcium is associated with changes in the pattern of protein phosphorylation within the egg.

In the ooplasm, sperm nuclear decondensation continues and the nuclear envelope disperses. Decondensation of the sperm nucleus is caused by a decrease in the number of disulfide cross-linkages within the chromatin. The expanded sperm chromatin becomes reconstituted as the **male pronucleus** identifiable by the reformation of the nuclear envelope as a series of flattened vesicles followed by the appearance of a variable number of dense nucleoli. Very little is known about the formation of the **female pronucleus**. After extrusion of the second polar body, the chromatin of the ovum becomes diffuse and is subject to protein modification. The nuclear membrane then reappears marking the end of the formation of the pronucleus. As the pronuclei approach each other, chromatin duplication takes place in both pronuclei.

GAMETE TRANSPORT AND FERTILIZATION

As prophase of the first cleavage division is approached, chromosomes condense and the pronuclear membranes disappear. Two sets of condensed chromosomes come together and arrange themselves on the equatorial plane. The sister chromatids then separate and migrate in opposite directions. The nuclear membrane reforms and a cleavage furrow can be observed. The two blastomeres mark the end of the first cleavage division.

Polyspermy is fertilization of an ovum by multiple sperm cells. Mammalian eggs establish blocks to polyspermy at the level of the zona pellucida and vitelline membrane. In response to the calcium signals initiated by fusion of the egg and sperm cell membranes, exocytosis of granules from the cortex of the egg modifies the sperm attaching receptors on the zona pellucida, thus preventing sperm cell attachment. The zona pellucida becomes resistant to protein digestion by cross-linking the tyrosine residues. At the molecular level, little is known about the oolemma block to polyspermy. Changes in protein composition of the oolemma have been noted between fertilized and nonfertilized mammalian eggs. In many invertebrate and nonmammalian species, a change in membrane potential of the oolemma serves as a block to polyspermy. However, no changes in the membrane potential of fertilized mammalian eggs have been detected.

Oviductal Secretions and their Role in Fertilization

Oviductal fluid is a selective transudate of the blood containing secretions of the oviduct epithelium. Oviductal fluid contains numerous ions, proteins and growth factors. **Oviductin**, an estrogen associated glycoprotein, is in highest concentration in oviductal fluid around the time of ovulation. Oviductin binds to the sperm surface and is responsible in maintaining motility and viability. Oviductin is also involved in stimulating capacitation of spermatozoa. Oviductin binds to the zona pellucida and the oolemma of the ovum. Oviductin reduces the rate of sperm binding causing an increase in the fertilization rate and development potential of the ovum. Oviductin also enhances cleavage of the fertilized ovum thus facilitating embryonic development.

Specificity to Fertilization

Behavioral and physiological mechanisms are in place to insure those members of a species breed and fertilize with other members of the same species. Below is a description of some of the physiological mechanisms that ensures species specificity to fertilization.

Fertilization specificity can be expressed by the failure of spermatozoa of one species to be transported to the site of fertilization in the female of another species. Often spermatozoa placed in the female reproductive tract of a foreign species survive poorly when compared to native spermatozoa. However, spermatozoa that do survive in the female reproductive tract of a foreign species can capacitate and fertilize the eggs of their own species, although they cannot penetrate the zone pellucida surrounding the eggs of that host species.

A specificity of complementary receptors on the surface of the gametes can play an important part in the maintenance of the reproductive isolation between species. It appears in many mammalian species that capacitated spermatozoa will not bind to the receptors on the zona pellucida of a foreign egg. The preeminence of the zona pellucida over the oolemma as prime controller of interspecies fertilization is clearly evident when the zona pellucida is removed. Hamster eggs in which the zona pellucida has been removed will fuse and incorporate the spermatozoa of many other species.

There appears to be no specificity when it comes to the interaction between sperm chromatin and ooplasmic factors needed for pronucleus formation. It has been shown that hamster ooplasm will transform

GAMETE TRANSPORT AND FERTILIZATION

the chromatin of a human spermatozoon into a pronucleus. This piece of knowledge may be used to develop techniques to examine the chromosomal constitution of human spermatozoa.

<u>Errors in Fertilization</u>

During the process of fertilization, events can happen that brings abnormal chromosome numbers to the cells of developing embryos. Some of these conditions are described below.

Polyspermy is a rare condition in which an ovum is fertilized by more than one sperm cell. These embryos usually develop normally for a short period of time before degenerating. Aging and heating the ovum prior to fertilization can enhance this condition. Therefore, polyspermy is more likely to occur in females that are bred too late or have elevated body temperatures.

Other rare errors in fertilization include polygny, gynogenesis and androgenesis. **Polygny** occurs when a spermatozoon fertilizes an ovum that has failed to extrude the polar body during meiosis. **Gynogenesis** is a condition in which the male pronucleus fails to develop, and all the genetic material in the fertilized egg is female. **Androgenesis** is a condition in which the male contributes all the genetic material.

Since females are born with the total number of eggs that they will ever have, aging of the eggs occurs with the aging of the female. In an ovum that is stimulated to complete meiosis later in life, there is a greater chance of **nondisjunction**. Such failures result in an extra chromosome in the female pronucleus that should have passed to the first polar body prior to ovulation. Fertilization of such an egg creates a condition called **trisomy.** An example of a trisomic condition in which development often continues to birth, is **Down's syndrome** [John Down, English physician, 1828-1896]. These individuals have an abnormal appearance and an impaired mentality.

6

MAMMALIAN EMBRYOLOGY

EARLY EMBRYONIC DEVELOPMENT IN MAMMALS

<u>Human Embryonic Development Prior to Uterine Implantation</u>

Embryology is the science that deals with the origin and development of an individual organism. After fertilization in the human, the **zygote** (diploid cell resulting from the fusion of the sperm cell and an ovum) slowly moves down the oviduct in the direction of the uterus. Passage of the conceptus through the oviduct is accomplished by ciliary activity, assisted by the relative absence of mucous. During this passage, the human **embryo** (the developing human to approximately the 8th week of gestation) is going through a series of mitotic divisions. An embryo of 8 or more cells confined by the zona pellucida is referred to a **morula** (Figure 6.1). By three days post fertilization, the embryo is a mass of 32 cells.

The human embryo enters the uterus approximately 3 to 4 days after fertilization. The human embryo appears as a large mass of 30 to 200 cells and is now referred to as the **blastocyst** (Figure 6.1). From the outside, the blastocyst looks like a solid ball of cells confined within the zona pellucida. Examination of the blastocyst reveals a fluid filled cavity referred to as the **blastoceal**. A single layer of cells just beneath the zona pellucida of the blastocyst is called the **trophoblast**. A clump of cells underneath the trophoblast at one end of the blastocyst is called the **embryoblast** (inner cell mass). The trophoblast surrounds the blastoceal and the embryoblast. The embryoblast eventually gives rise to the tissues and organs of the embryo.

The blastocyst rests freely within the uterine cavity for about 2 to 3 days. It derives its nutrients from the fluid secreted from the uterine glands. The human endometrium is approximately 10 mm to 14 mm thick, and has been prepared for embryonic invasion by a complex activity of cytokines, growth factors and lipids. Receptivity of the human endometrium to invasion by the embryo is between day 16 to day 22 of the menstrual cycle. Approximately six days after fertilization, the uterus secretes a proteolytic enzyme that dissolves the zona pellucida surrounding the blastocyst. With the removal of the zona pellucida, the embryoblast adheres to the upper posterior endometrial wall along the midsaggital plane. Attachment of the blastocyst to the endometrium involves adhesion molecules including cytokines and integrins.

<u>Uterine Implantation</u>

Implantation is the attachment of the blastocyst to the uterine endometrium. **Invasive implantation** is the invasion of the uterine endometrium by the embryo. Invasive implantation is particularly strong in primates and rodents (Figure 6.2). Connective tissue cells of the uterine endometrium undergo a **decidual reaction.** During the decidual reaction, endometrial connective tissue cells at the implantation site become loaded with glycogen and lipids and assume a polyhedral shape. The trophoblast of the human blastocyst secretes collagenases, gelatinases and stomelysins that degrade the extracellular material between the epithelial cells of the uterine endometrium. The epithelial cells of the endometrium are lifted off the basement membrane allowing the embryonic trophoblast cells to interdigitate between and beneath them. Destroyed uterine cells release lipids, carbohydrates, nucleic acids and proteins that are taken up by the conceptus. The trophoblast adjacent to the embryoblast differentiates into the syncytiotrophoblast and cytotrophoblast. The **syncytiotrophoblast** (a synctium) is the outer layer of the trophoblast and appears as

MAMMALIAN EMBRYOLOGY

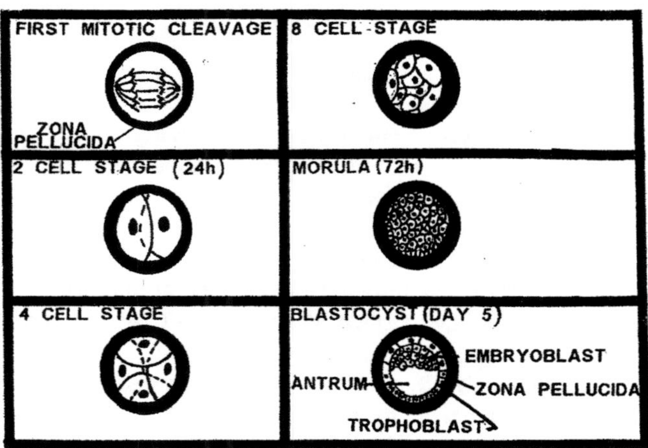

Figure 6.1.   First five days of human embryogenesis.

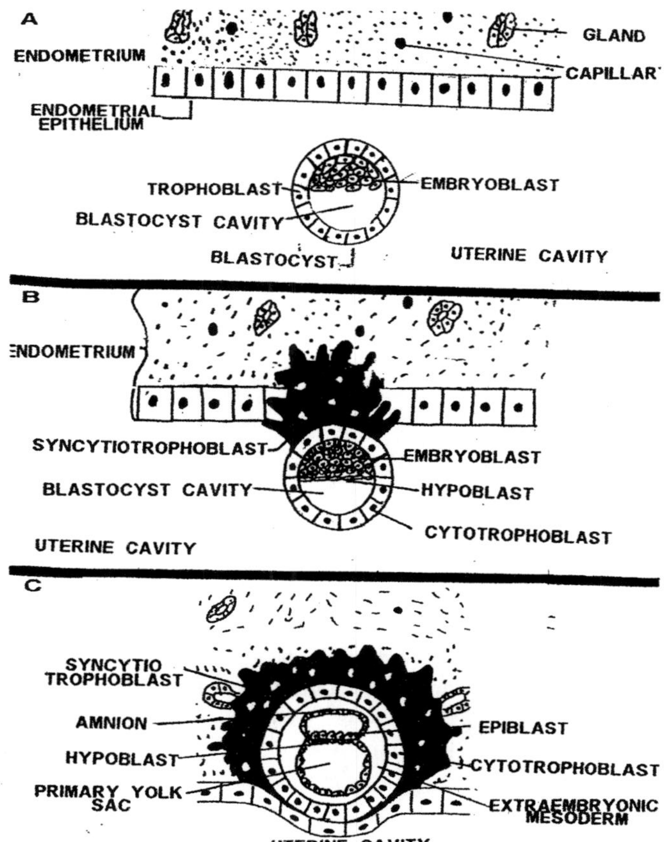

Figure 6.2.   Invasive implantation of the human embryo. (A) Approchment of the blastocyst to the uterine lining.
   (B) Partially implanted blastocyt 8 days after fertilization (C) Implantation of the blastocyst to the uterine
   lining 10 days after fertilization.

a multinucleated mass without distinct cell boundaries. The syncytrophoblast invades the uterine endometrium in species with invasive implantation The **cytotrophoblast** is the inner mononucleated layer of cells.

The depth of conceptus invasion of uterine tissue is species specific. In **interstitial implantation** (e.g. human), the conceptus invades the uterine stroma so deeply that the surface epithelium is restored over it. Trophoblastic cells of the embryo invade the uterine endometrium by sending out cellular projections that become the placental villi. Blood capillaries grow into these cords from the vascular system of the embryo. By the sixteenth day of gestation, blood is flowing through of the placental villi. Simultaneously, blood sinuses from the mother surround the placental villi. In **eccentric implantation**, the uterine stroma is only partially invaded and the conceptus projects into the uterine cavity. Secondary contact may occur with the uterine lining on the opposite side of the uterus resulting in two sites of placental development (e.g. rhesus monkey) or a complete belt of placenta around the waste of the conceptus (e.g. dog and cat).

Noninvasive implantation is that attachment of the conceptus to the endometrial surface without invasion. The conceptus of a noninvasive species is supported for a longer period of time on uterine secretions (when compared to invasive species) prior to implantation. This allows the conceptus of noninvasive implantation species to grow more extraembryonic tissue (conceptus grows to a bigger preimplantation size) thus establishing a vast surface area where the exchange of metabolites between the uterus and developing conceptus can occur. In the horse and pig, attachment of the conceptus occurs at multiple sites over the external surface of the conceptus. In ruminants, attachment of the conceptus is limited to distinct areas of projecting aglandular mucosae (**uterine caruncles**). Between sites of attachment between the conceptus and uterine tissue, uterine glands continue to secrete nutrients for the conceptus.

Delayed implantation is the ability to suspend the blastocyst *in utero* prior to implantation. In the rat and mouse, the blastocyst can be suspended in the uterine cavity for several days prior to implantation. **Facultative delayed implantation** occurs naturally in the female mouse and rat due to the suppressed secretion of endogenous estrogens in response to suckling by the previous litter. Animal species displaying **obligatory delayed implantation** (e.g. roe-deer, badger, elephant seal, fruit bat, mink and brown bear) have a conceptus (es) suspended in the uterine cavity for a period of weeks or months prior to implantation. This allows mating to occur when adults are in prime condition and allows the offspring to be born when nutritional conditions are optimal.

<u>Second Week of Human Embryonic Development</u>

The rapid proliferation and differentiation of the trophoblast characterizes the second week of human gestation (Figure 6.3). The **primary yolk sac** develops and the **extraembryonic mesoderm** arises from the endoderm of the primary yolk sac. The **extraembryonic coelom** forms from cavities that develop in the extraembryonic mesoderm. A **secondary yolk sac** develops as the primary yolk sac regresses. The **amnionic cavity** appears as a space between the cytotrophoblast and the embryoblast. The embryoblast differentiates into a **bilaminar embryonic disc** containing the **epiblast** (high columnar cells next to the amnionic cavity) and the **hypoblast** (a layer of small cuboidal cells facing the blastocystic cavity). The **prechordal plate** develops and becomes the future site of the mouth and an important organizer of the head region.

The amnion, allantois and chorion are membranes that develop around the embryo during early gestation. The **amnion** is a membrane that surrounds the amnionic cavity. The amnion is formed when the cells of the epiblast separate and organize themselves around the amnionic cavity. The **allantois** is a membrane that extends from the caudal wall of the yolk sac into the connecting stalk. The allantois can be seen in the human embryo as early as day 16 post-fertilization. In some primitive mammals (e.g. platypus), the allantois is large (compared to the human embryo) and has a respiratory function as well as acting as a

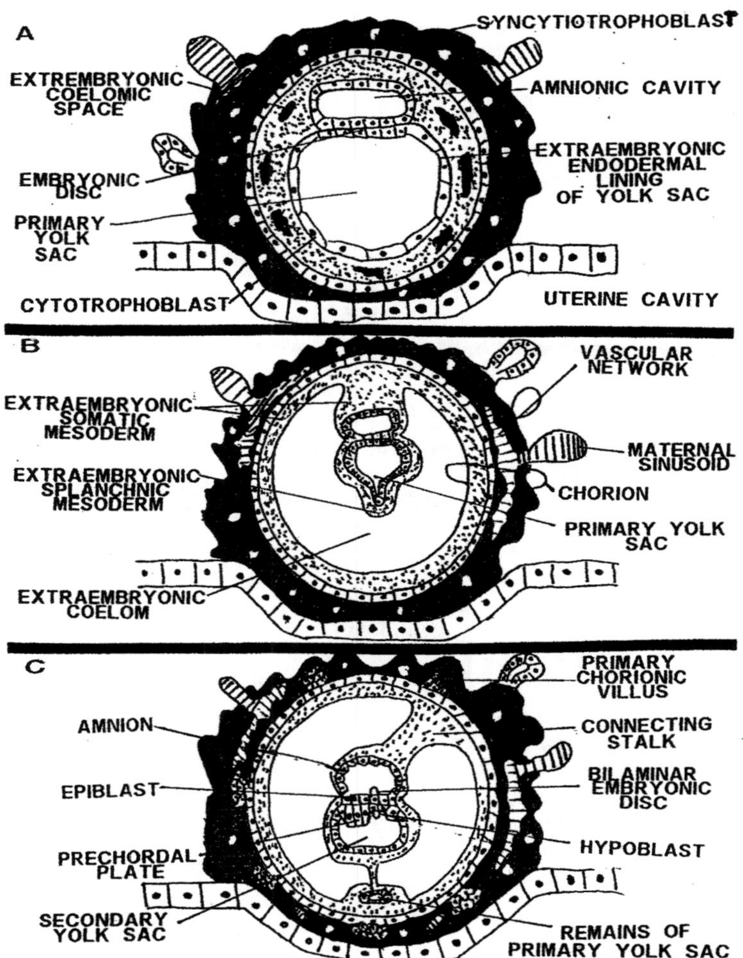

Figure 6.3. The second week of human embryogenesis. (A) Twelve days after fertilization. (B) Thirteen days after fertilization. © Fourteen days after fertilization.

reservoir for urine during development. The allantois of the human embryo is involved in early blood cell formation and is involved in the development of the urinary bladder. The allantois also forms the **urachus** during development. As development proceeds, the blood vessels of the allantois eventually form the umbilical arteries and veins. The human umbilical ligament is formed from the allantois.

In early human embryogenesis, the extraembryonic coelom splits the extraembryonic mesoderm into the **extrasplanchnic mesoderm** (surrounds the yolk sac) and the **extraembryonic somatic mesoderm** (covers the amnion and trophoblast). Two layers of trophoblast and the extraembryonic somatic mesoderm form the **chorion**. The chorion forms the walls of the chorionic sac. The extraembryonic coelom is now called the **chorionic cavity.**

<u>Third Week of Human Embryonic Development</u>

A localized thickening of the epiblast develops into the **primitive streak** (Figure 6.4). **Mesenchymal cells**

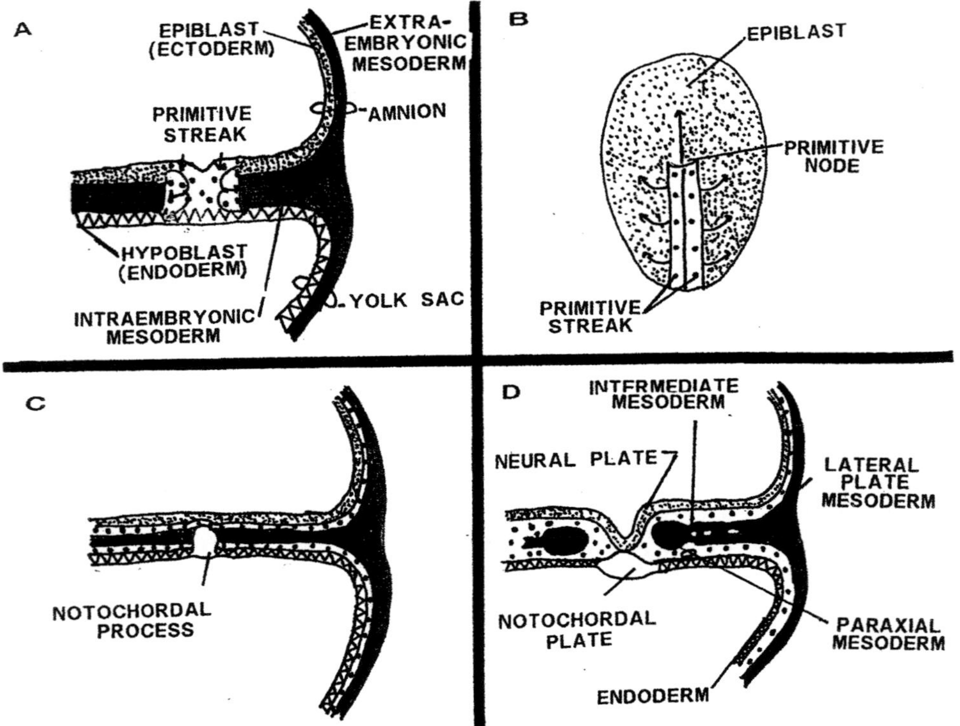

Figure 6.4. Third week of human embryogenesis. (A). Transverse section of embryonic disc 16 days after fertilization. (B). Dorsal view of embryonic disc 16 days after fertilization. (C). Transverse section of presomite embryo 18 days after fertilization. (D). Transverse section of an embryo with one pair of somites 20 days after fertilization.

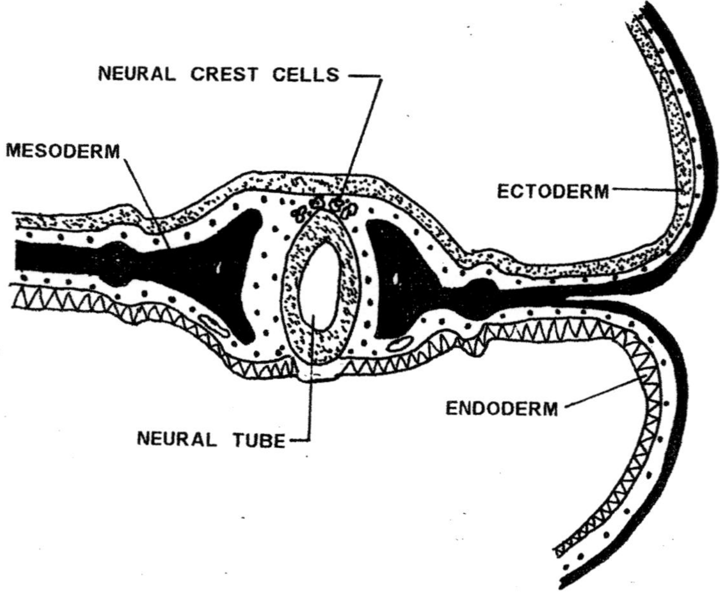

Figure 6.5. Human embryogenesis 22 days after fertilization. Transverse section through the fourth somite of a seven somite embryo.

MAMMALIAN EMBRYOLOGY

(fibroblast-like cells) derived from the invagination of epiblastic cells of the primitive streak migrate between the epiblast and the hypoblast. As soon as the primitive streak starts to produce mesenchymal cells, the epiblast becomes known as the **embryonic ectoderm**. The **embryonic endoderm** is formed by displacement of the hypoblast by cells of the epiblast. The intraembryonic mesoderm is formed by the organization of the mesenchymal cells. The **embryonic mesoderm** exists between the ectoderm and the ectoderm by the end of the third week of gestation. Mesenchymal cells arising from the **primitive node** of the primitive streak form the **notochordal process**. The **notochordal canal** is formed by the extension of the primitive pit into the notochordal process. Coalescence of openings in the floor of the notochordal canal forms the **notochordal plate**. The **notochord** (the primitive axis of the embryo around which the axial skeleton forms) is formed by the infolding of the notochordal plate.

Neurulation is the process of forming the neural plate and neural tube (Figure 6.5) The **neural plate** appears as a thickening of the embryonic ectoderm cranial to the primitive node. The **neural groove** (a longitudinal groove in the neural plate) develops and is flanked by **neural folds**. The **neural tube** forms by fusion of the neural folds. **Neural crest** formation occurs between the surface ectoderm and the neural tube. Formation of the neural crest involves the migration of neuroectodermal cells dorsilaterally from the neural folds to the area between the surface ectoderm and the forming neural tube. Sensory ganglia of the cranial and spinal nerves are formed by the neural crest. Other neural crest cells migrate to form other aspects of the nervous system.

Paraxial mesoderm is formed by the thickening of the mesoderm on the lateral sides of the notochord. **Somites**, compact aggregates of mesenchymal cells, form by division of the paraxial columns. Cells of the somites migrate to form the vertebrae, ribs and axial musculature. **Coelomic vesicles** (in the lateral mesoderm) and cardiogenic mesoderm coalesce to form the **coelom** (kidney to horseshoe-shaped cavity). The coelom eventually gives rise to the body cavities of the organism.

Blood vessels first develop on the yolk sac, allantois and chorion, and then on the embryo. **Blood islands** (spaces) appear in the aggregations of mesenchyme. Blood islands become lined by endothelium of mesenchyme origin and eventually unite to form the **primordial cardiovascular system**. Paired endothelial tubes combine the form a tubular heart by the end of the third week of gestation. The primitive heart is connected to blood vessels of the embryo, chorion and yolk sac. Endothelial cells of blood vessels of the yolk sac and chorion give rise to **hemangioblasts** (primitive blood cells).

Primary chorionic villi acquire mesenchymal cores and are transformed into **secondary chorionic villi**. **Tertiary chorionic villi** develop when capillaries develop in the secondary chorionic villi. Cytotrophoblastic extensions from these stem villi join to form the cytotrophoblastic shell that anchors the chorionic sac to the uterine endometrium. Rapid development of the chorionic villi increases the surface area for the exchange of nutrient and waste material between the embryo and the mother.

<u>Classification of Placental Forms</u>

Mammalian placentas can be histologically classified based on the number of tissue layers separating the fetal blood supply from the maternal blood supply. These different placental types are discussed below (Figure 6.6).

A **hemochorial placenta** has 3 distinct layers of fetal tissue separating the fetal blood supply from the maternal blood sinus. A **hemonochorial placenta** (e.g. humans and rhesus monkeys) has a single layer of trophoblast between the fetal blood supply and the maternal blood supply. The shape of human placental attachment is **discoid**, whereas that of the rhesus monkey is **bidiscoid**. A **hemodichorial placenta** (e.g.

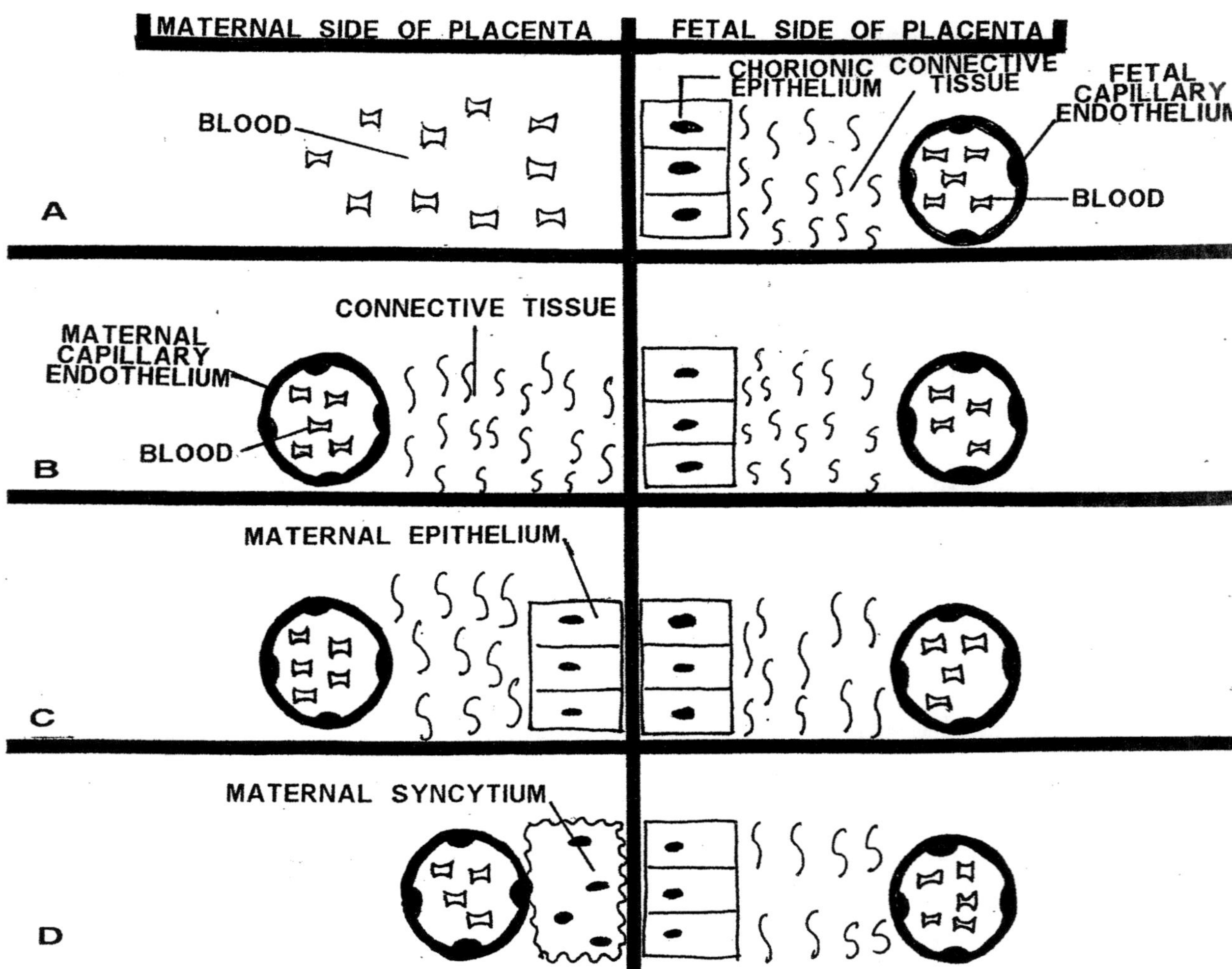

Figure 6.6. Histological classification of placental types. (A) Hemochorial placenta. (B). Endotheliochorial placenta. © Epitheliochorial placenta. (D). Synepitheliochorial placenta.

rabbit) has two layers of trophoblast between the fetal blood supply and the maternal blood supply. The shape of the placental attachment is discoid. A **hemotrichorial** placenta (e.g. rat and mouse) have three layers of trophoblast between the fetal blood supply and the maternal blood supply. The shape of the placental attachment is discoid.

An **endotheliochorial placenta** is a placenta in which chorionic tissue penetrates to the endothelium of the maternal blood vessel (e.g. dog, cat, bear and mink). Gross examination of placentas obtained from dogs and cats reveals that the chorionic villi form an equatorial zone around the placenta. This type of placentation is described as **zonary**.

MAMMALIAN EMBRYOLOGY

An **epitheliochorial placenta** is a placenta in which the chorionic tissue is in contact (but does not erode) the uterine endometrium (e.g. pig, mare, whales, lemurs, dolphins, deer and giraffe). The **epitheliochorial placenta** has six layers of tissue separating the fetal blood supply from the maternal blood supply. In pigs and horses, the placental villi develop over the entire surface of the placenta. This placental type is described as **diffuse**.

A **synepitheliochorial placenta** is a placenta in which the epithelial cells of the endometrium fuse with binucleated cells from the fetal side to form a synctium (e.g. sheep and cattle). Placental villi arise in tufts that are distributed over the chorion. This type of placentation is described as **cotyledonary** or **multiplex**.

The type of placentation can influence passive immunity from the mother to the newborn. In species with epitheliochorial placentation, there are too many layers of tissue separating the fetal blood supply from the maternal blood supply to allow the transfer of antibodies from mother to fetus. In species with hemochorial placentation, the number of tissue layers separating the fetal blood supply from the maternal blood supply is reduced enough to allow antibodies to cross from the mother to the fetus.

Function of the placenta

The placenta has numerous functions during mammalian gestation. The placenta is involved in the transfer of oxygen, nutrients, gamma globulins and waste material between mother and the fetus. Transfer of gases, nutrients and waste material is primarily by diffusion. The exchange of gamma globulins between the mother and fetus involves specific carrier mechanisms. The mammalian placenta stores glycogen, produces hormones, is involved in the initiation of labor, and is involved in the development and onset of lactation. The placenta serves as a protective barrier preventing the passage of most macromolecules, bacteria and numerous other agents between mother and fetus. However, the placenta does not block the passage of many drugs, chemicals, vitamins and antibodies that typically evoke maternal immune responses.

Evolution of Placental Forms

The production of gametes by meiosis creates offspring with greater genetic variation. This provides greater flexibility for species entering new environments and/or responding to changes in their existing environments. Genetic variation is often observed in the degree of protein polymorphism. For example, the molecular structure of serum albumin is very similar between most mammalian classes, suggesting that polymorphism for albumin has been selected against. However, gamma globulins can show a considerable degree of polymorphism from one generation to the next within a mammalian species.

The evolution of viviparity brought another challenge to organisms that produced gametes by meiosis. Viviparity limits the amount of variation tolerated between generations because of the degree of intimacy between the fetal and maternal circulatory systems. However, as previously discussed, the exogenous environment favors genetic variability between generations of organisms. A compromise between these conflicting interests is to delay the production of polymorphic proteins between the fetus and the mother until after the birth. This results in a longer period of infancy in mammalian species when compared to other animal species.

Throughout mammalian evolution, mammals have entered new environments and have been exposed to rapidly changing environmental conditions. Many eutherian mammalian species have epitheliochorial placentation. This type of placentation allows for greater genetic variability to be expressed *in utero* between generations and an earlier development of body systems prior to birth. Because of the thick barrier separating maternal from fetal blood supplies, birth occurs before the maternal immune system has time to recognize fetal foreign antigens and develop antibodies against them.

An increase in the size and structure of the brain has been a critical component in the environmental adaptation and evolution of primates. The epitheliochorial placenta, efficient in the transportation of nutrients from the mother to the fetus, is not as efficient in the transport of oxygen from maternal to fetal blood. A reduction in the number of placental tissue layers between maternal and fetal blood supply was required to improve oxygen transport. This gave rise to the hemochorial placenta. Five million years ago, ancestral hominids were exposed to further environmental changes. The hemochorial placenta, that had been important in earlier primate evolution, became disadvantageous when genetic variability (the production of polymorphic proteins) between hominid generations was needed to respond to environmental challenges. This problem was remedied by the delayed production of polymorphic proteins until after birth. This explains the many underdeveloped human body systems at the time of birth and increased period of infancy when compared to other primate species.

EMBRYONIC ORGANOGENESIS

<u>Introduction</u>

Organogenesis is the development of organs during embryonic development. Human organogenesis takes place approximately from the 4th to the 8th week of gestation. During this period, the lungs, pancreas, liver, and digestive tract all develop from the primitive gut. The beginnings of the urogenital system, muscular system, skeletal system, and nervous system are also being established. Blood circulation is initiated and the heart is beating by approximately day 23 post-fertilization. As previously discussed, the embryoblast gives rise to three layers of tissue called the endoderm, mesoderm and the ectoderm. The endoderm (the inner most layer) forms the lining of the gut, glands and the urinary bladder. The ectoderm (the outer most layer) forms the elongated ridge of the central axis of the germ disc. The mesoderm (located between the endoderm and the mesoderm) gives rise to the connective tissues, vasculature, bones, muscles and the adrenal cortex. The outer layer of the mesoderm differentiates into three regions. The first mesodermal region develops into the vertebrae that encases the neural tube. The second mesodermal region forms the skeletal muscles. The third mesodermal region forms the connective tissue of the skin. A general discussion of the organogenesis occurring in major body systems is discussed below.

<u>Embryonic Development of the Pharyngeal Apparatus</u>

The **pharyngeal apparatus** is an early embryonic structure that consists of arches, pouches, grooves and membranes (Figure 6.7). The pharyngeal apparatus is responsible for the development of the jaw, tongue, palate, pharynx, lips, face and neck. Each **pharyngeal arch** is composed of a core of mesenchyme (mesodermal origin) surrounded internally by endoderm and externally by ectoderm. As embryogenesis proceeds, neural crest cells migrate into the arches and are a source of the various connective tissue components in the facial and oral regions. Each pharyngeal arch is composed of a muscular component, a nerve, an artery, and a rod of cartilage. **Pharyngeal grooves** externally separate the pharyngeal arches. **Pharyngeal pouches** are evaginations of the pharynx that internally separate the arches. **Pharyngeal membranes** are formed when the endoderm of the pouches contacts the ectoderm of the grooves.

The epithelial lining of the external auditory meatus is derived from the first pair of pharyngeal grooves. The second, third and fourth pairs of pharyngeal grooves disappear during embryogenesis. The tympanic cavity, mastoid antrum and Eustachian tube [Bartolomo Eustachio, Italian anatomist, 1524 – 1574] are derived from the first pharyngeal pouch. The palatine tonsil is derived from the second pharyngeal pouch.

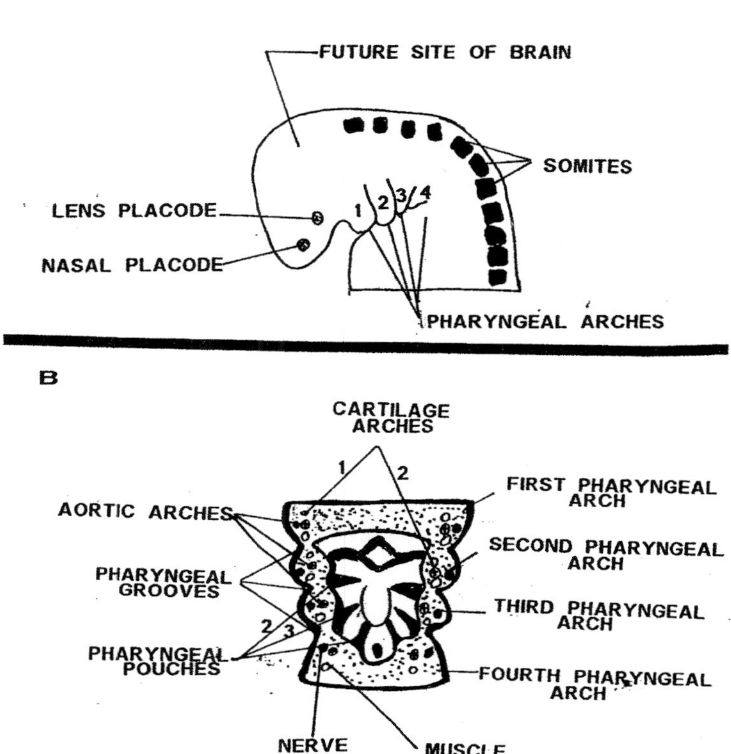

Figure 6.7. The human pharyngeal apparatus approximately 28 days after fertilization. (A) Location of the pharyngeal apparatus in relation to the head, neck and thorax of the embryo. (B) Horizontal section through the embryo showing the floor of primordial pharynx.

The third pair of pharyngeal pouches gives rise to the thymus. The parathyroid glands are formed from the third and fourth pairs of pharyngeal pouches. The first pair of pharyngeal membranes gives rise to the tympanic membrane. All other pharyngeal membranes disappear during embryogenesis.

Embryonic Development of the Cardiovascular System

Development of the human cardiovascular system is observable by the third week of gestation. The heart starts to beat at approximately day 23 post-fertilization. **Heart tubes** develop from isolated cells clusters formed from the proliferation of mesenchymal cells derived from the splanchnic mesoderm. The heart tubes fuse to form the primordial heart (Figure 6.8). The primordial heart consists of the **sinus venosus, atrium, ventricle** and the **bulbus cordis**. The **truncus arteriosus** is continuous with the bulbus cordis. The bulbus cordis eventually becomes part of the ventricles. The heart becomes four chambered between the 4th and 7th week of gestation. The vitelline, cardinal, and umbilical systems of veins empty into the developing heart. The **umbilical systems of veins** involute after birth of the fetus. The **embryonic vitelline vein system** gives rise the portal system of veins. The **embryonic cardinal vein system** develops into the caval vein system. **Aortic arches** are derived from the **embryonic aortic sac**. The aortic arches

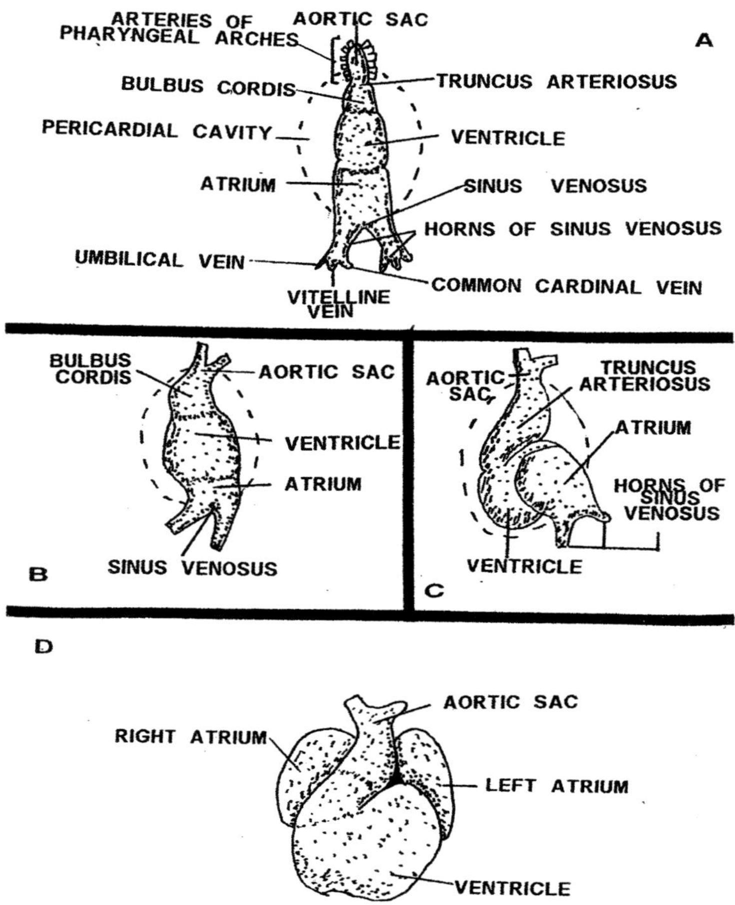

Figure 6.8. Embryonic development of the human heart. (A) Primordial heart approximately 24 days after fertilization. (B) and (C) Bending of the primordial heart to the right in response to accelerated growth (compared to the rest of the heart) of the bulbus cordis and ventricle. (D) View of the embryonic heart approximately 35 days after fertilization.

invade the pharyngeal arches at approximately the 4th to 5th week of gestation. Transformation of the aortic arches into the carotid, subclavian and pulmonary arteries occurs during the 6th to 8th week of gestation.

Development of the embryonic vascular system requires **vasculogenesis** (formation of the vascular system) and **angiogenesis** (development of new blood vessels). Blood vessel formation starts on extraembryonic membranes before being observed occurring in the body of the embryo. Blood vessels are observed on the chorion as early as day 12 post-fertilization. Mesenchymal cells differentiate into vessel forming cells called **angioblasts**. Angioblasts aggregate to form isolated angiogenic cell clusters called **blood islands** (Figure 6.9). Small cavities appear within the blood islands by the confluence of the interstitial spaces. Angioblasts form the primitive endothelium by arranging themselves around the cavities of the blood islands. These endothelium-lined cavities then fuse to form networks of endothelial channels (vasculogenesis). Vessels sprout into adjacent areas by endothelial budding and fuse with other vessels

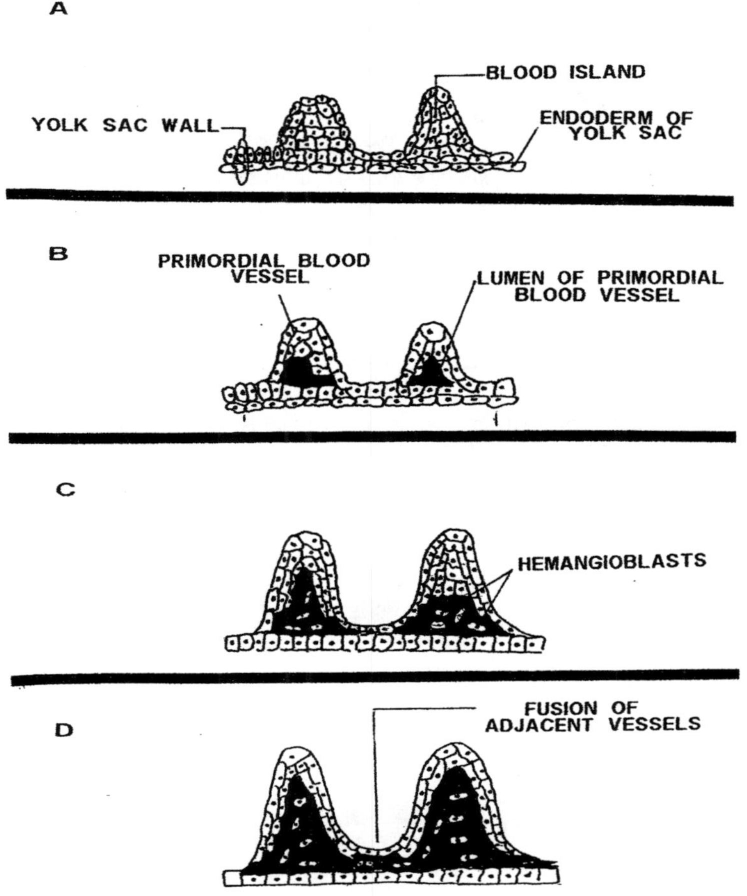

Figure 6.9. Human embryonic development of blood and blood vessels. (A) Formation of blood islands. (B) Appearance of cavities in blood islands. (C) Formation of vessel endothelium and hemangioblasts in blood island cavities. (D) Formation of endothelial channels.

(angiogenesis). Mesenchymal cells surrounding the primordial endothelial blood vessels differentiate into the muscular and connective tissue layers.

The formation of **hemangioblasts** (primitive blood cells) is observed by the end of the 3rd week of gestation. As blood vessels develop on the yolk sac and allantois, hemangioblasts develop from the endothelial cells of blood vessels. Hemopoiesis does not begin in the embryo until the 5th week of gestation. As gestation continues, hemopoiesis occurs in various parts of the embryonic mesenchyme starting first in the liver and later in the spleen, bone marrow and lymph nodes.

<u>Embryonic Development of the Nervous System</u>

As previously discussed folds of the neural plate give rise to the neural tube. From the rostral end of the developing neural tube three primary vesicles known as the **prosencephalon** (forebrain), **mesencephalon** (midbrain) and **rhombencephalon** (hindbrain) are observed (Figure 6.10). The prosencephalon divides into the **telencephalon** and **diencephalon.** The rhombencephalon divides into the **metencephalon** and **myelencephalon**. The mesencephalon remains undivided. The telencephalon gives rise to the cerebrum, basal ganglia, olfactory bulbs and hippocampus. The diencephalon gives rise to the epithalamus, thalamus, hypothalamus, pituitary, pineal body, optic nerve and mammillary bodies. The metencephalon gives rise to the pons and cerebellum. The medulla oblongata develops from the myelencephalon. The dura mater develops from the mesoderm that surrounds the neural tube. The pia mater and arachnoid mater develops from the cells of the neural crest. The neurohypophysis develops from evagination of the neuroectoderm of the diencephalon. The adenohypohysis develops from evagination of ectoderm from the roof of the primitive mouth (Rathke's pouch [Martin Rathke, German anatomist, physiologist, and pathologist, 1793 – 1860]).

The sympathetic and parasympathetic branches of the autonomic nervous system originate from the basal plate of the neural tube and cells of the neural crest (Figure 6.5). Sympathetic preganglionic neurons within the intermediolateral column develop from neuroblasts of the neural tube. Sympathetic postganglionic neurons with sympathetic chain ganglia and prevertebral ganglia originate from neural crest cells. Parasympathetic preganglionic neurons within the nuclei of the midbrain, pons and spinal cord (levels S_2 to S_4) also develop from neuroblasts of the neural tube. Parasympathetic postganglionic neurons within the ciliary, pteryopalatine, submandibular and enteric ganglia develop from neural crest cells. Postganglionic neurons within ganglia of the abdominal and pelvic cavities also develop from neural crest cells. **Glioblasts** give rise to supporting cells of the central nervous system. These support cells would include astrocytes, oligodendrocytes, ependymocytes, tanycytes, choriod plexus cells and microglia cells.

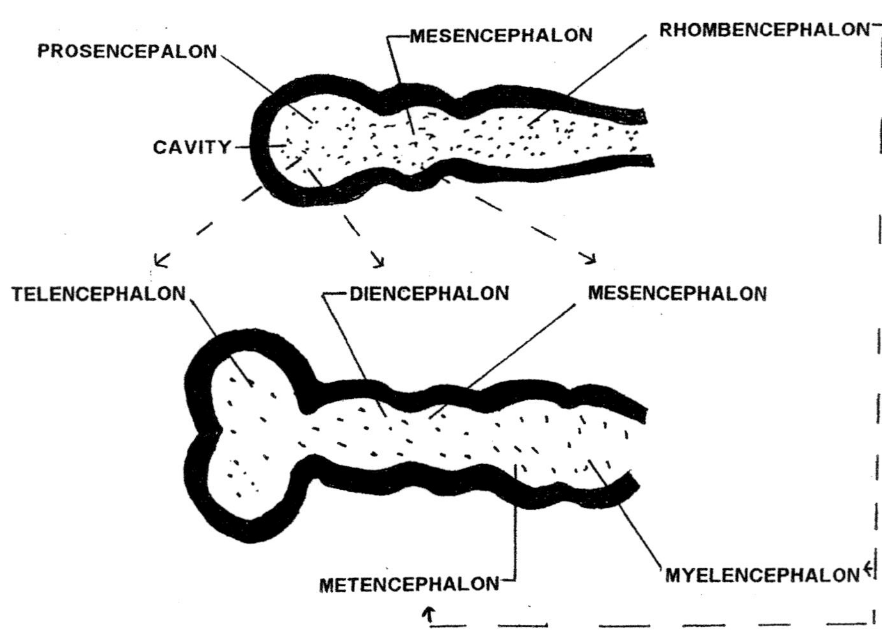

Figure 6.10. Embryonic development of the human brain. The development of the brain secondary vesicles is completed by the 5th week of gestation.

MAMMALIAN EMBRYOLOGY

The eye is formed partially from a neuroectodermal evagination of the diencephalon called the **optic cup** and **optic stalk**. The retina, iris and ciliary body are derived from the optic cup (Figure 6.11). The optic nerve develops from the optic stalk (Figure 6.12). The lens and the anterior epithelium of the cornea develop from the **lens placode** (surface ectoderm). The sclera, substantia propria of the cornea, corneal epithelium, vitreous body and extraocular muscles are derived from the embryonic mesoderm. The choriod coat and ciliary muscles are derivatives of neural crest cells. The embryonic **hyaloid artery** and **hyaloid vein** develop into the central artery and central vein of the retina, respectively.

Human **otic placodes** and the **otic vesicle** develop from embryonic surface ectoderm during the 4th week of gestation. The internal ear develops from the otic vesicle (Figure 6.13). As embryogenesis proceeds, the otic vesicle divides into the **utricular portion** and the **saccular portion.** The utricle, semicircular ducts, and vestibular ganglion develop from the utricular portion of the otic vesicle. The saccule, cochlear duct, organ of Corti [Alfonso Corti, Italian anatomist, 1822 – 1876] and spiral ganglion of the 8th cranial nerve develop from the saccular portion of the otic vesicle. The middle ear develops from embryonic pharyngeal arches, pouches and membranes. The ossicles of the middle ear, tensor tympani muscle and stapedius muscle is derived from the first and second pharyngeal arches. The epithelial lining of the auditory tube and middle ear cavity are derived from the first pharyngeal pouch. The tympanic membrane is derived from the first pharyngeal membrane. The epithelial lining of the external auditory meatus and the auricle of the external ear are derived from the first pharyngeal groove and auricular hillocks.

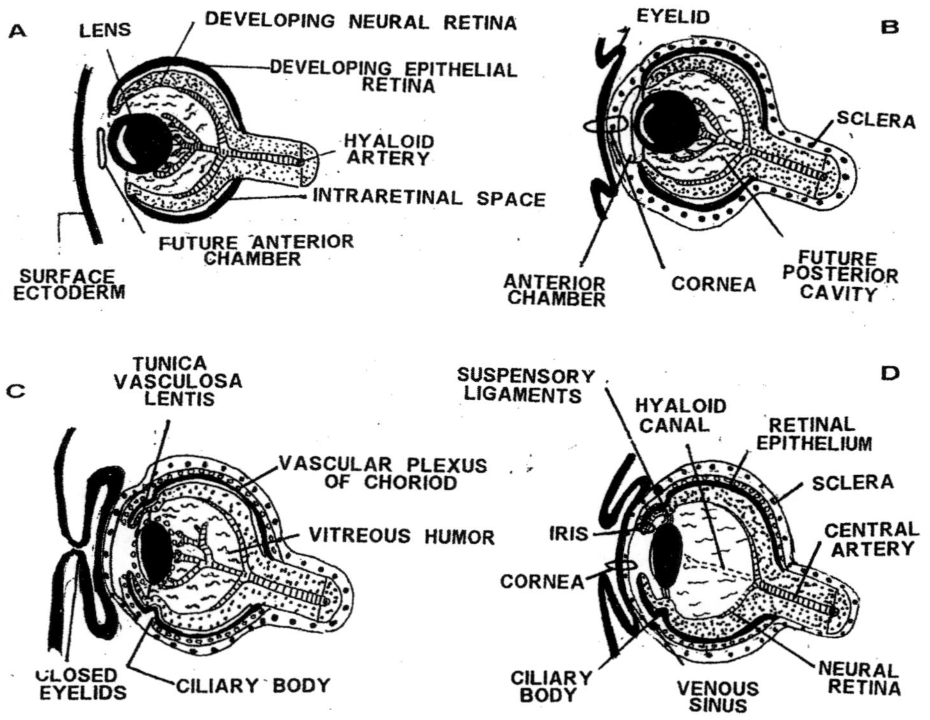

Figure 6.11. Development of the human eye. (A) Eye development at 5 weeks post-fertilization. (B) Eye development at 6 weeks post-fertilization. © Eye development at 20 weeks post-fertilization. (D) Anatomy of the eye after birth.

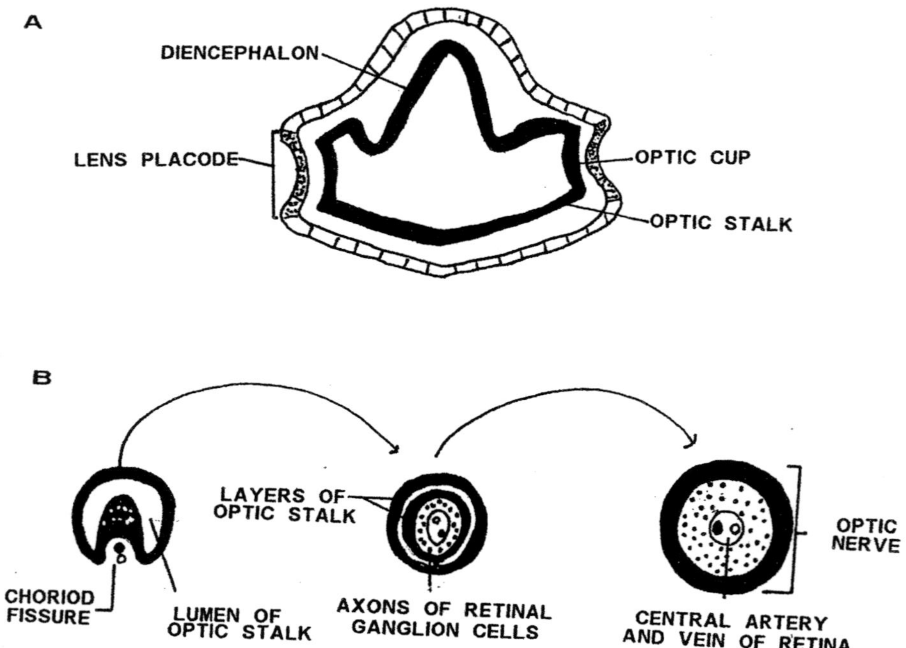

Figure 6.12. Development of the optic nerve. (A) Diencephalon showing the optic cup and optic stalk. (B) Formation of the optic nerve from the optic stalk.

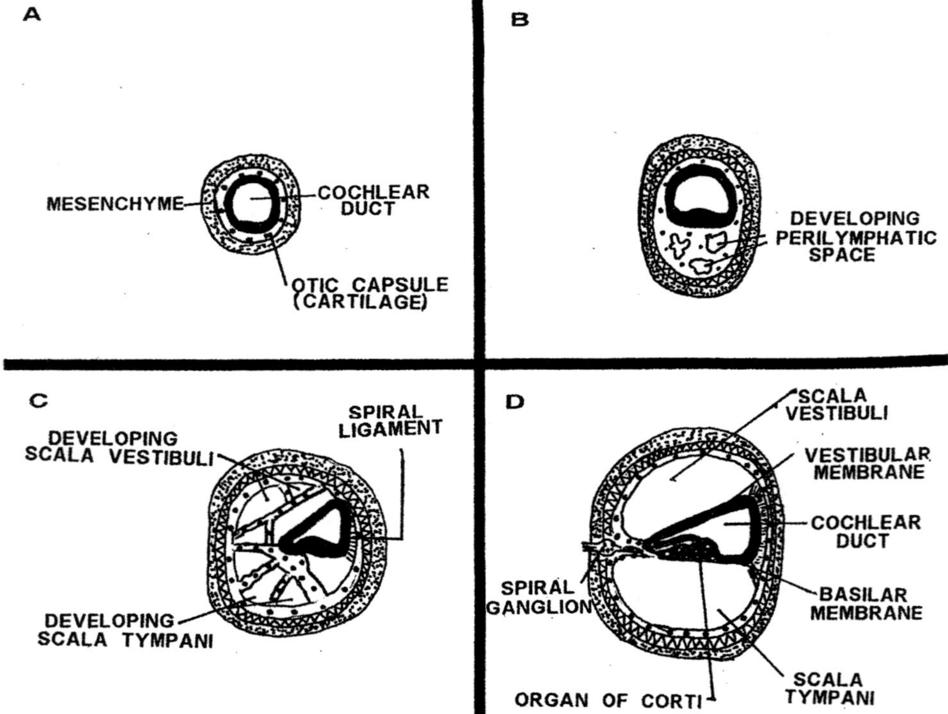

Figure 6.13. Development of the human internal ear. Transverse sections (A – D) showing cochlear development from the 8th week to the 20th week of gestation.

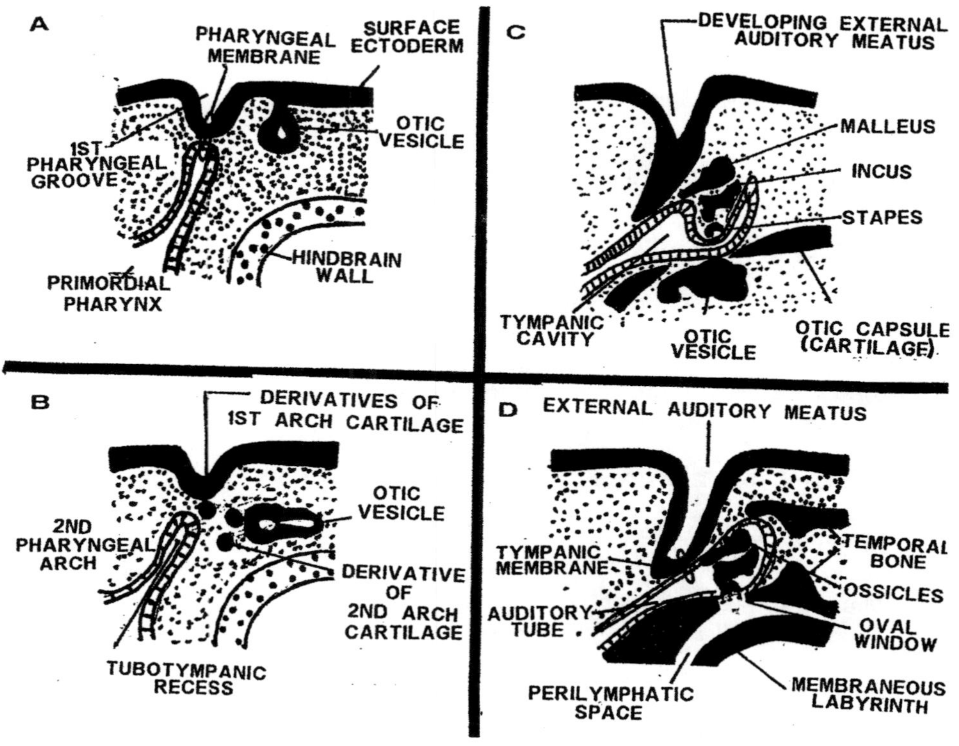

Figure 6.14. Development of the human external and middle ear. (A) Ear anatomy at the 4th week of gestation showing the otic vesicle and the pharyngeal apparatus. (B) Ear anatomy showing tubotympanic recess and pharyngeal arch cartilages at the 5th week of gestation. (C) Tubotympanic recess enclosing the ossicles of the middle ear. (D) Final stage of ear development.

Embryonic Development of the Digestive System

The human **primitive gut tube** is formed by the incorporation of the yolk sac into the embryo during craniosacral and lateral folding of the embryo. The epithelial lining and glands are derived from the endoderm. The mucosa, submucosa, muscularis and serosa are derived from the mesoderm. The epithelial lining of the gut proliferates rapidly and obliterates the lumen before canalization occurs. The primitive gut tube is suspended in the peritoneal cavity of the embryo by ventral and dorsal mesenteries that develop into adult mesenteries. The primitive gut tube can be divided into the foregut, midgut and hindgut. Parts of the primitive gut and their derivatives are discussed below (Figure 6.15).

The esophagus, stomach, liver, gallbladder, bile duct, pancreas and the upper duodenum are derivatives of the **foregut**. The esophagus is formed when the **tracheoesophageal septum** divides the foregut into the esophagus and trachea. The stomach develops from a fusiform dilation in the foregut during early

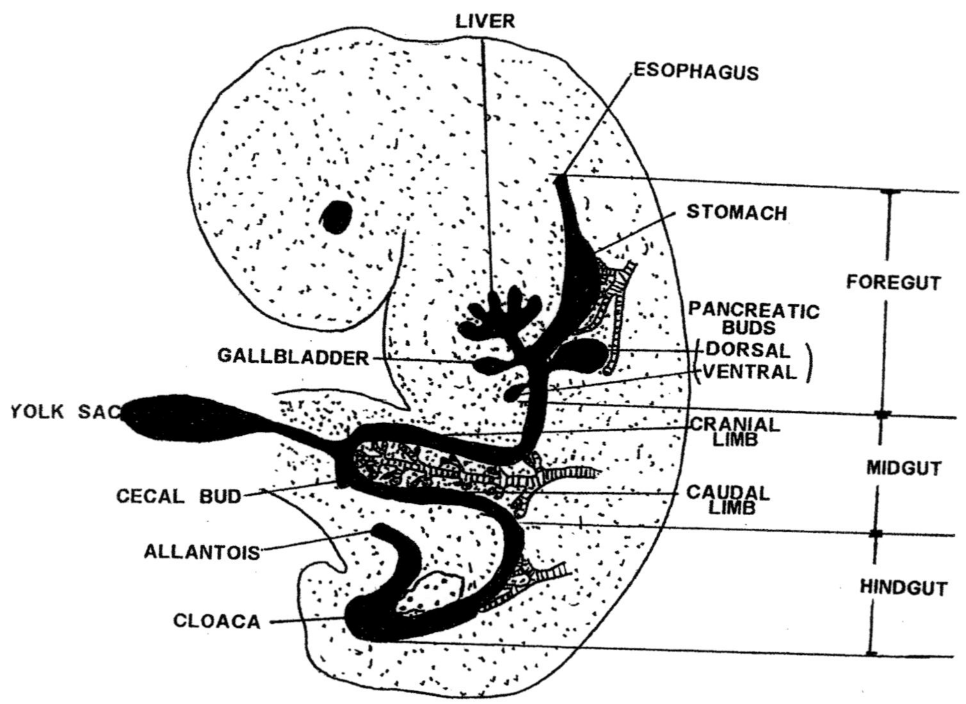

Figure 6.15. Development of the human digestive tract.

organogenesis. A 90-degree rotation of the primitive stomach during its formation gives rise to the lesser peritoneal sac. The liver is an endodermal outgrowth of the foregut. The **hepatic diverticulum** (of mesodermal origin) sends out hepatic cord cells that surround vitelline veins forming the **hepatic sinuses**. As organogenesis proceeds, the connection between the hepatic diverticulum and the foregut narrows forming the bile duct. An outgrowth of the bile duct gives rise to the gallbladder and cystic duct. **Pancreatic buds** give rise to the pancreas. The **ventral pancreatic bud** forms the uncinate process and a portion of the pancreatic head. The dorsal pancreatic bud forms the remaining part of the pancreatic head, body and tail. The cranial part of the duodenum develops from the caudal portion of the foregut.

The lower duodenum, jejunum, ileum, cecum, appendix, ascending colon and two-thirds of the transverse colon are derivatives of the **midgut**. The lower duodenum is derived from the **cranial portion of the midgut**. During **umbilical herniation** (midgut loop herniates through the umbilicus), the cranial limb of the midgut loop forms the jejunum and cranial portion of the ileum. The caudal portion of the ileum, cecum, appendix, ascending colon and the proximal two-thirds of the transverse colon are derived from **caudal limb of the midgut loop**.

MAMMALIAN EMBRYOLOGY

The distal one-third of the transverse colon, descending colon, sigmoid colon, upper anal canal and urogenital sinus are derivatives of the **cranial end of the hindgut**. The lower anal canal develops from invagination of the surface ectoderm.

<u>Embryonic Development of the Urinary System and Adrenal Glands</u>

The **urogenital ridge** (of mesodermal origin) develops along the dorsal body wall of the embryo. The **nephrogenic cord** (a portion of the urogenital ridge) forms the pronephros, mesonephros and metanephros. The **pronephros** completely regresses as embryonic development proceeds. The **mesonephros** forms the **mesonephric (Wolffian) duct** [Kaspar F. Wolff, German embryologist in Russia, 1733 – 1794]. The **metanephros** eventually develops into the kidney.

The metanephros begins to develop into the permanent human kidney at approximately the 5th week of gestation (Figure 6.16). The **metanephric diverticulum**, an outgrowth of the **metanephric duct,** penetrates the **metanephric mass of the intermediate mesoderm** and serves as the primordia for the ureter, renal pelvis, calyces and collecting ducts. The stalk of the metanephric diverticulum gives rise to the ureter while the cranial end develops into the renal pelvis. Straight collecting ducts branch forming generations of collecting tubules. The first four generations of collecting tubules coelesce to form the major calyces of the kidney while the second four generations of collecting tubules coelesce to form the minor calyces. The ends of the arched collecting tubules induce clusters of mesenchymal cells to form the **metanephric vesicles** (Figure 6.17). The metanephric vesicles elongate to form the **metanephric tubules** that eventually develop into the nephrons of the kidney. Glomeruli invaginate into the proximal end of the metanephric tubules whereas the distal end attaches to the collecting tubules. As gestation continues, the kidney ascends from the sacral region of the body to its adult location.

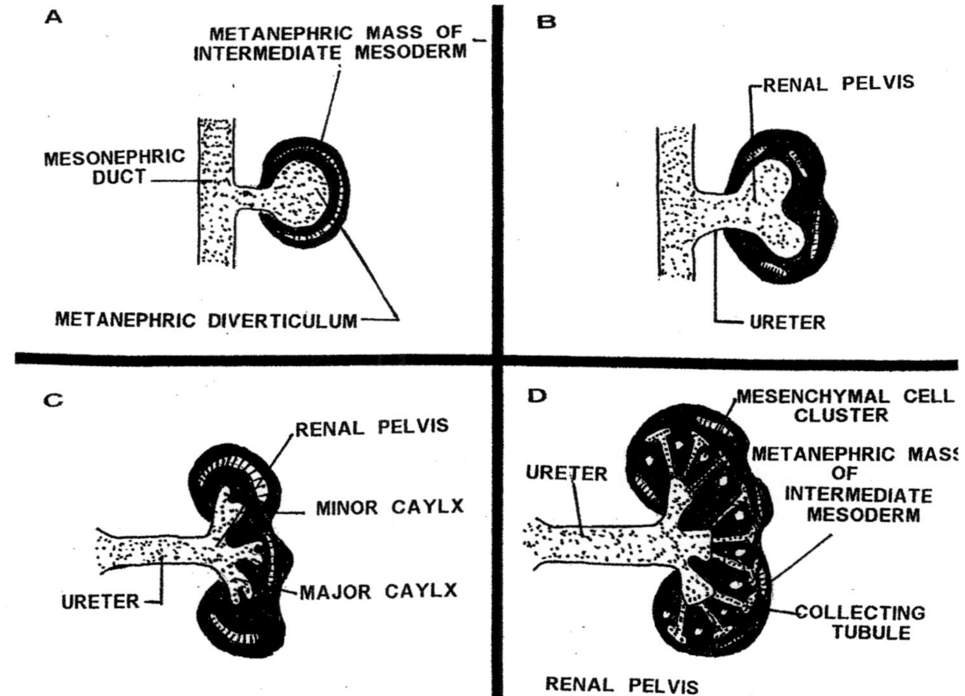

Figure 6.16. Development of the metanephros of a human embryonic kidney (A) Anatomy of the metanephros at 5 weeks post-fertilization. Continued development of the metanephos from the 5th to the 8th week of gestation (B – D).

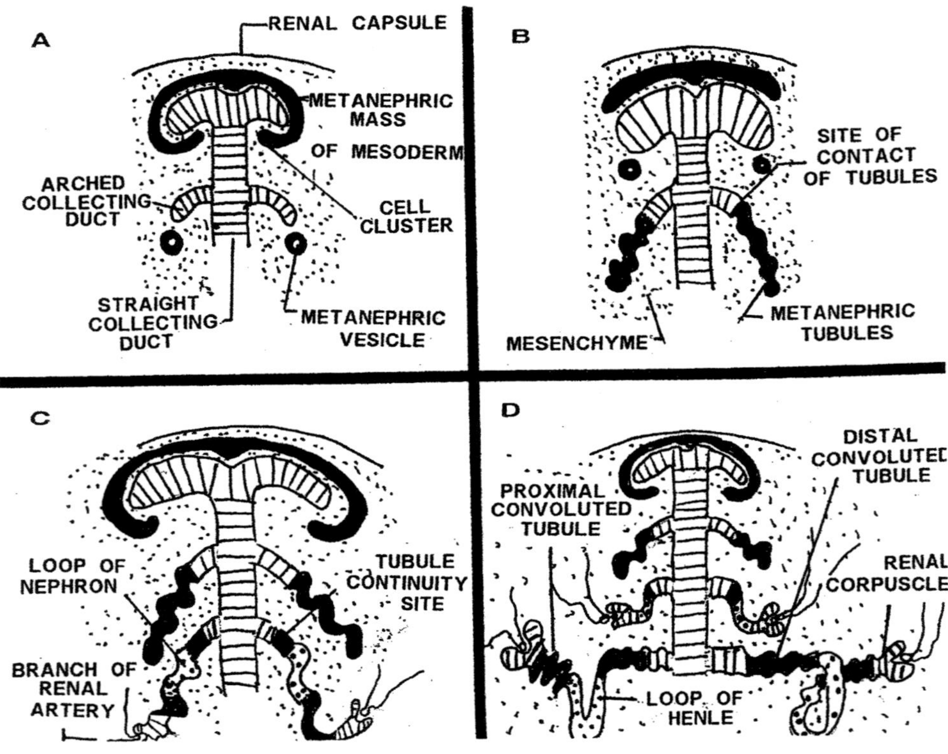

Figure 6.17. Nephrogenesis in the developing human kidney. (A) The initiation of renal nephrogenesis in the human embryo at the 8th week of gestation. The continuation of nephrogenesis (B – D).

Development of the human adrenal gland starts at approximately the 6th week of gestation and is completed at approximately 4 years post-partum (Figure 6.18). Development of the fetal adrenal cortex can be observed starting at approximately the 6th week of gestation. Cells of the fetal adrenal cortex are derived from the mesothelium of the abdominal wall. During the 8th week of gestation, the mesenchymal cells from the mesothelium begin to enclose the fetal cortex eventually giving rise to the permanent cortex of the adrenal gland. Differentiation of the permanent adrenal cortex into zones is initiated during late fetal development. The zona glomerulosa and the zona fasciculata are present in the permanent adrenal cortex at birth, whereas the zona reticularis is not observable until 3 years post-partum. The medulla of the developing adrenal gland is derived from cells of an adjacent sympathetic ganglion. These cells eventually develop into the secretory cells of the medulla. During gestation, the medulla remains relatively small when compared to the cortical area of the developing kidney. The size of the adrenal gland decreases considerably after birth. Although the adrenal medulla increases markedly in size after birth, regression of the fetal cortex is primarily responsible for the decrease in adrenal size.

The urinary bladder develops from the upper end of the urogenital sinus that is continuous with the allantois. During embryogenesis the allantois degenerates forming the **urachus**. The trigone of the bladder is formed from the incorporation of the lower end of the mesonephric duct into the posterior wall of the urogenital sinus.

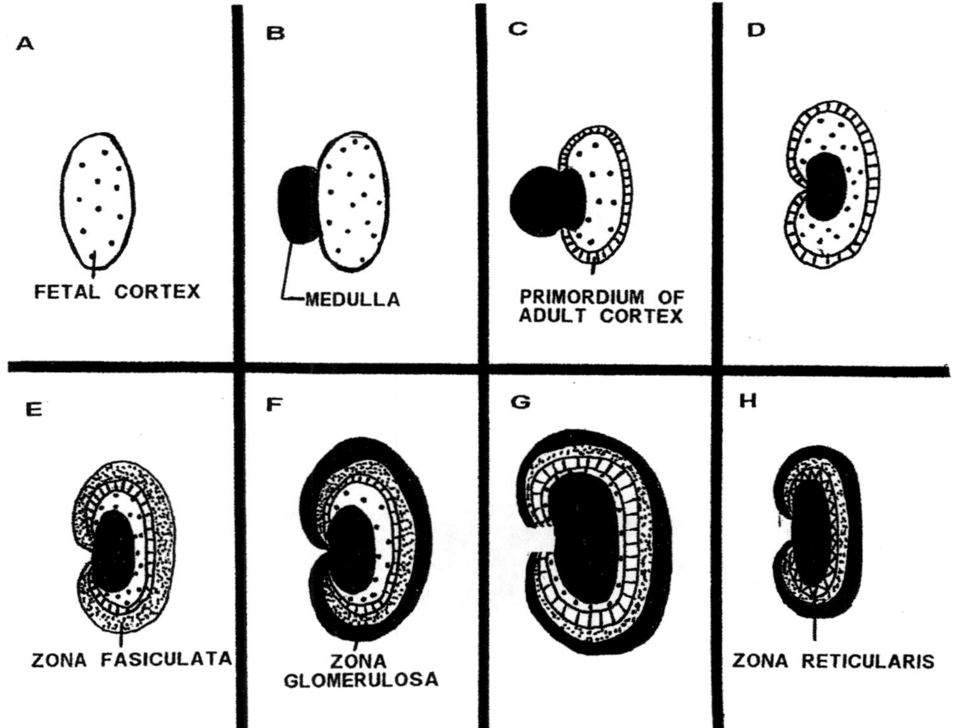

Figure 6.18. Development of the human adrenal gland. Adrenal gland at the 6th week of gestation (A), 7th week of gestation (B), 8th week of gestation (C) fetal development (D and E), neonatal period (F), 12 months post-partum (G) and 4 years of age (H).

<u>Embryonic Development of the Reproductive System</u>

The gender of human embryos is indistinguishable during the first 6 weeks of gestation. By week 12, some characteristics of the male and female external genitalia are recognizable. By the 20th week of gestation, components of the indifferent embryo are remodeled to form the reproductive systems of the male and female.

The female reproductive system develops from several parts of the indifferent embryo (Figure 6.19). The ovaries develop from the **indifferent gonads** of the embryo. The **paramesonephric duct** gives rise to the oviducts, uterus, cervix and the superior end of the vagina. The urinary bladder, urethra, the urethral and paraurethral glands, greater vestibular glands and the inferior portion of the vagina develop from the **urogenital sinus** of the indifferent embryo. The **phallus** of the indifferent embryo gives rise to the clitoris. The labia minora develops from the urogenital folds of the indifferent embryo. The **labioscrotal swellings** of the embryo give rise to the labia majora of the female.

The testes of the male develop from the **indifferent gonads** of the embryo (Figure 6.20). The

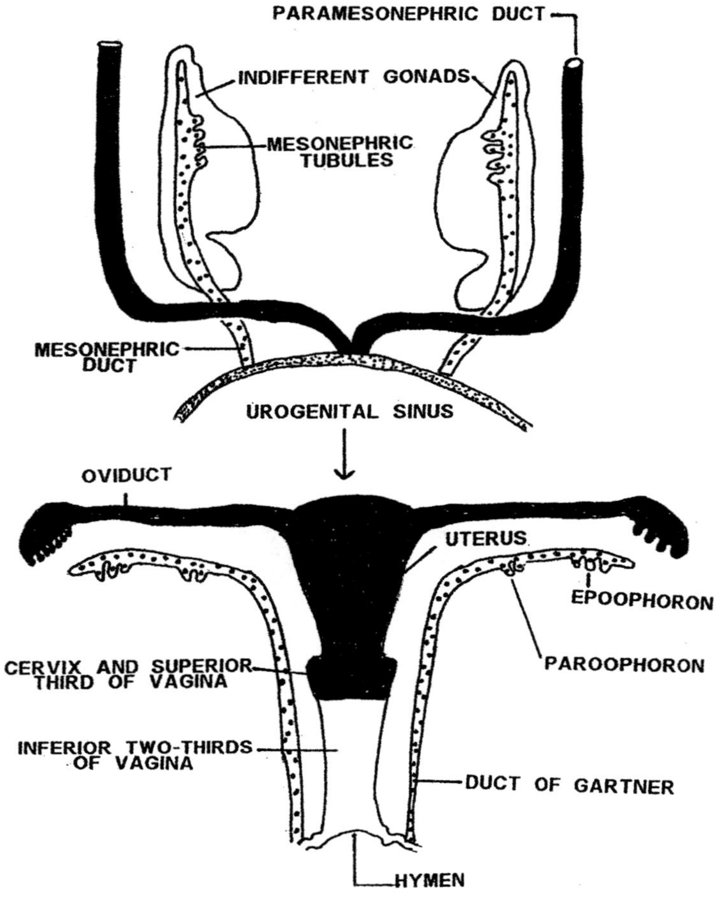

Figure 6. 19. Embryonic development of the human female reproductive tract.

mesonephric duct of the indifferent embryo gives rise to the epididymis, ductus deferens, seminal vesicles and the ejaculation ducts of the male. The efferent ducts of the male are derived from the **mesonephric tubules** of the indifferent embryo. The **urogenital sinus** of the indifferent embryo gives rise to the urinary bladder, urethra, prostate and the bulbourethral glands. The penis is derived from the **phallus** of the indifferent embryo. The **urogenital folds** of the indifferent embryo give rise to ventral aspect of the penis and penile raphe. The scrotum and scrotal raphe develops from the **labioscrotal swellings** of the indifferent embryo.

The indifferent embryo will develop into a female unless masculation occurs. **Testes-determining factor (TDF)**, **mullerian inhibiting factor (MIF)** and testosterone (produced by the fetal testes) direct the indifferent embryo towards development as a male.

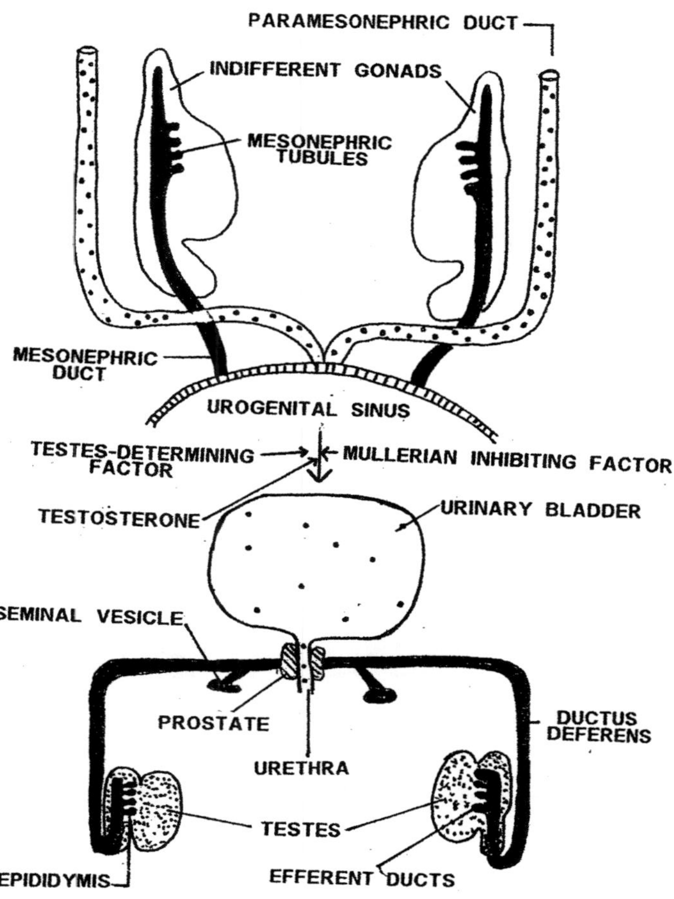

Figure 6.20. Embryonic development of the human male reproductive tract.

<u>Embryonic Development of the Lower Respiratory Tract</u>

The lower respiratory tract is made of the trachea and lungs. The lower respiratory system begins to develop at approximately the 4th week of gestation (Figure 6.21). The lower respiratory system starts to develop from the **median laryngeal groove** on the floor of the **primordial pharynx**. The groove deepens to form the **laryngotracheal diverticulum**. The laryngotracheal diverticulum is separated from the foregut by the **tracheoesophageal folds** that develop into the **tracheoesophageal septum**. Epithelium of the lower respiratory structures and the tracheobronchial glands are derived from the **laryngotracheal tube**. The blood and lymphatic vessels, cartilage, connective tissue and muscle of the lower respiratory tract develop from the splanchnic mesoderm surrounding the laryngotracheal tube. The connective tissue of the larynx and the epiglottis are derived from the mesenchyme of the **pharyngeal arch**. The mesenchyme of the caudal pharyngeal arches give rise to the laryngeal muscles, whereas the laryngeal cartilage's are derived from the cartilaginous bars of the 4th and 6th pairs of pharyngeal arches.

The **tracheal bud** develops at the distal end of the of the laryngotracheal diverticulum. The tracheal bud

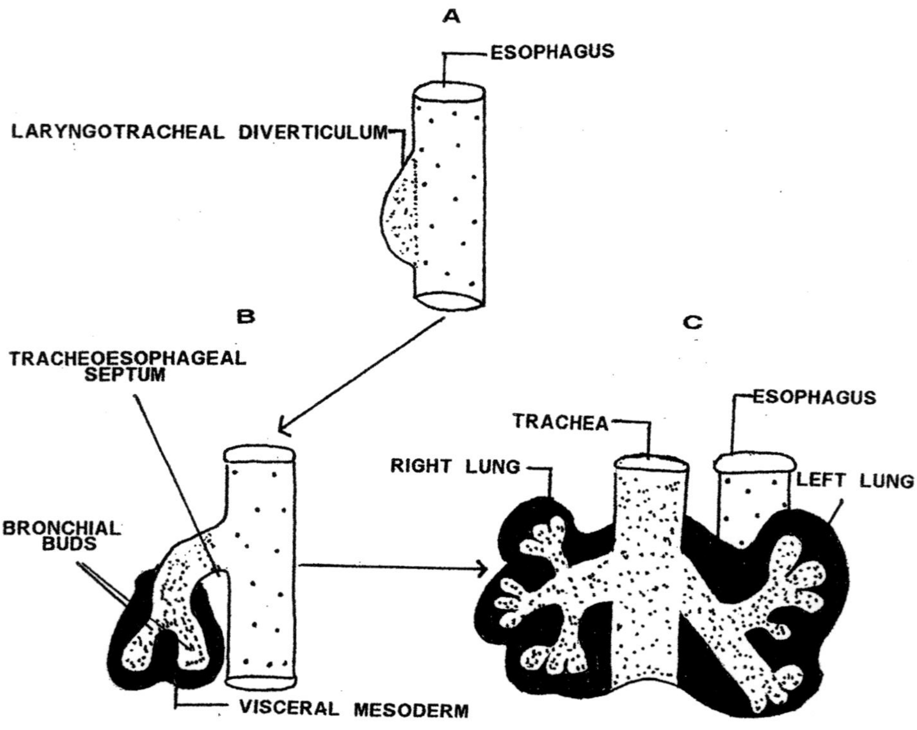

Figure 6.21. Embryonic development of the human respiratory system during the 4th week of gestation (A), the 5th week of gestation (B) and the 6th week of gestation (C).

develops into two **bronchial buds** at approximately the 5th week of gestation. The main bronchus for each lung is form by the enlargement of its respective bronchial bud. The bronchial buds for each respective lung gives rise to two new bronchial buds that later develop into secondary bronchi. The lobes of the developing embryonic lungs are supplied by their secondary bronchi. Progressive branching of each bronchus forms segmental bronchi. Each segmental bronchus with its surrounding mesenchyme is the primordium of the bronchopulmonary segment. Branching of airways continues in each lung through the rest of gestation and after birth until 23 generations are present in each lung.

7

MAMMALIAN FETAL DEVELOPMENT

MAMMALIAN FETAL GROWTH

<u>Fetal Growth during Gestation</u>

Fetal growth is the increase in mass of the conceptus from the end of the embryonic period to birth. Although fetal growth varies between different mammalian species, similar patterns can be observed. Fetal weight curves are generally sigmoid in shape and slow towards the end of gestation. This sigmoid pattern of growth is also seen in many organs that comprise the fetus (Figure 7.1). The decline in fetal growth weight prior to parturition appears to reflect a change from organ tissue aggregation to organ tissue differentiation in preparation for extrauterine life. Changes in the morphology and physiology of fetal organs (e.g. lung, liver and gut) are observed during this period of time. Changes in body composition also occur towards the end of gestation. Adipose tissue increases prior to birth in many mammalian species. This accumulation of adipose tissue towards the end of pregnancy varies widely between mammalian species ranging from 1% of the body weight in the rat to 16% of the body weight in humans. Body protein composition of the fetus at the end of pregnancy is similar between most mammalian species ranging from 12% to 14% of the body weight.

Factors influencing fetal growth are maternal, placental and fetal in origin. Maternal factors such as small maternal size, small uterine size, poor nutrition, inadequate uterine blood flow, anemia, tobacco smoking and substance abuse can drastically retard the growth of the fetus. Placental factors such as a small size, an inadequate density of villi, poor umbilical blood flow, inadequate hormone synthesis, poor nutrition production and utilization, and inadequate transporters and binding proteins can also be detrimental for the growth of the fetus. Fetal factors such as chromosome numbers, imprinted genes, nutrient production, feto-placental steroid production, fetal endocrine gland secretions as well as fetal growth factors can also influence fetal growth. Some of these factors are discussed below.

PLACENTAL PHYSIOLOGY DURING FETAL DEVELOPMENT

<u>Placental Blood Flow</u>

Blood flows into the placenta from the fetus and the mother (Figure 7.2). Fetal blood enters the placenta from the **umbilical arteries**. From the umbilical arteries, blood then flows into fetal capillaries of the placental villi. Fetal blood leaves the placenta when the fetal capillaries discharge their contents into the **umbilical veins**. Maternal blood enters the placenta through the **uterine arteries** and leaves the placenta via the **uterine veins**.

The direction of fetal and maternal blood flow can vary between different mammalian species (Figure 7.3). Placental blood flow between the mother and the fetus can be concurrent, crosscurrent, multivillous, or countercurrent. **Concurrent placental blood flow** is characterized by maternal and fetal blood flow being parallel and flowing in the same direction. In **countercurrent placental blood flow** (e.g. horse, guinea pig and rabbit), maternal and fetal blood flows are parallel to each other but in opposite directions.

107

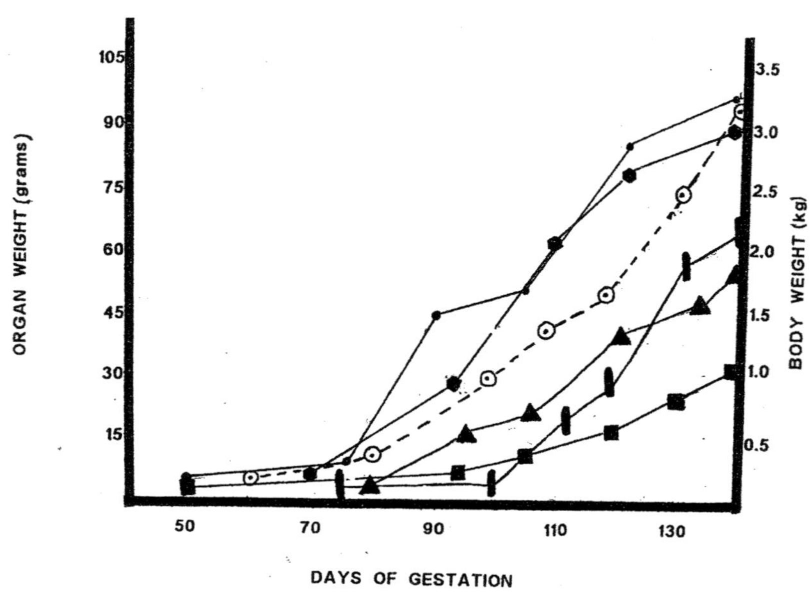

Figure 7.1. Increase in fetal body and selected organ weights in the lamb during gestation. Kidney (-■-), brain (-▲-), lungs (-●-), liver (-●-), gut (-■-) and body (-⊘-).

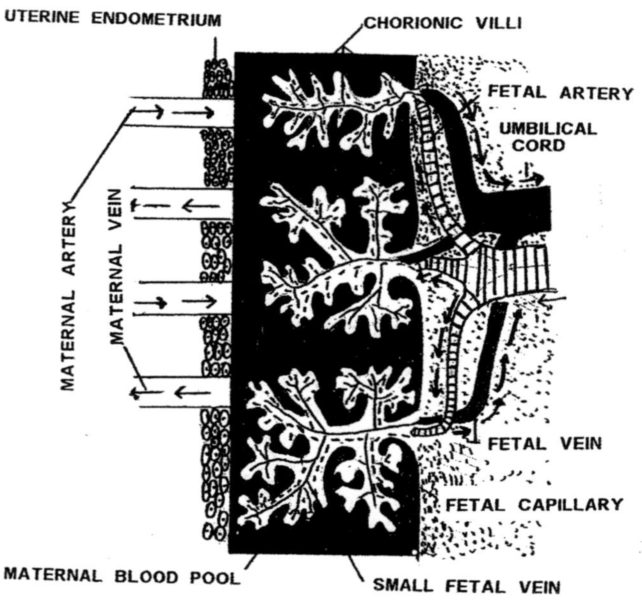

Figure 7.2. Maternal and fetal blood flow to and from the human placenta.

MAMMALIAN FETAL DEVELOPMENT

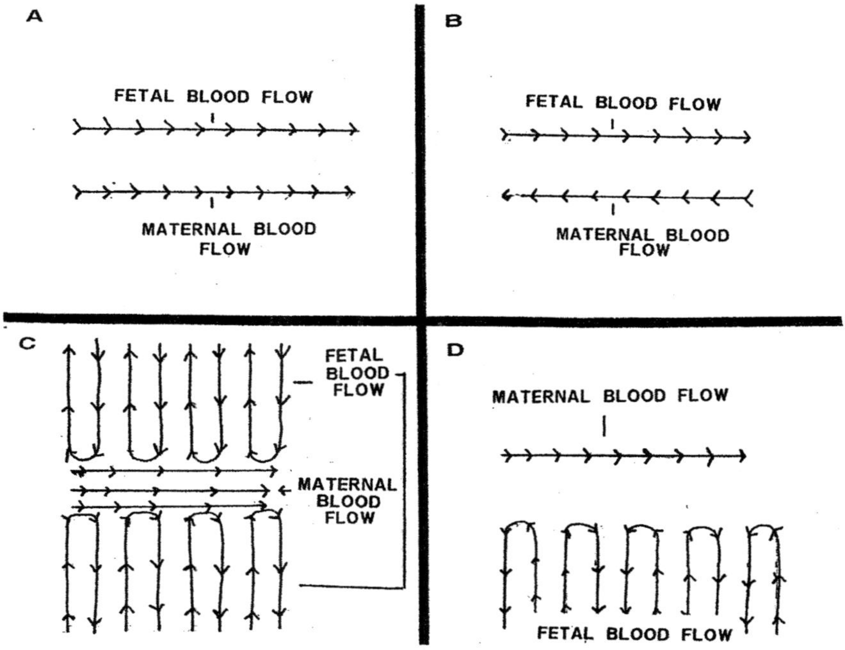

Figure 7.3. Fetal and maternal blood flow in the mammalian placenta. (A) Concurrent. (B) Countercurrent. (C) Multivillous (D) Crosscurrent.

Crosscurrent placental blood flow (e.g. sheep) is characterized by the maternal blood flowing perpendicular to that of the fetus. Maternal blood flowing between the placental villi where the fetal blood flows towards and then away from the tips of the placental villi characterizes the human placenta and is called **multivillous placental blood flow**. The different arrangements in placental blood flows influence the flow-limited transfer of substances between maternal and fetal blood.

Crosscurrent placental blood flow is the least efficient (when compared to countercurrent placental blood flow and multivillous placental blood flow) in the diffusion of oxygen from maternal blood to fetal blood. In sheep the partial pressure of oxygen in uterine venous blood is 57 mm Hg compared to 35 mm Hg in the fetal umbilical vein. Countercurrent placental blood flow is the most efficient in the diffusion of oxygen from maternal blood to fetal blood. In the horse, the partial pressure of oxygen in the uterine venous blood is 50 mm Hg compared to 49 mm Hg in the fetal umbilical vein thus demonstrating the efficiency of oxygen diffusion in countercurrent placental blood flow. Multivillous placental blood flow is slightly less efficient in the diffusion of oxygen from maternal blood to fetal blood. In humans, the partial pressure of oxygen in the uterine venous blood pool is 50 mm Hg compared to 41 mm Hg in blood contained in the fetal umbilical vein. Multivillous placental blood flow has both advantages and disadvantages compared to countercurrent placental blood flow. The disadvantage of multivillous placental blood flow is the loss of the efficient countercurrent exchange of flow-limited transfer of substances. Higher concentration gradients and special mechanisms of transport must be employed in multivillous placental blood flow to make up for this loss of efficiency. The advantage of multivillous placental blood flow is the tremendous increase in surface area over which maternal and fetal blood supplies are in contact for the diffusion of nutrients and gases.

MAMMALIAN FETAL DEVELOPMENT

<u>Placental Permeability and Membrane Diffusion Constant</u>

The placenta allows the diffusion of nutrients and wastes between the maternal and fetal vasculatures. The diffusion of material is dependent on placental membrane permeability and total membrane conductance. **Placental permeability** can be defined as the net quantity of substance that will diffuse for a given concentration difference. **Placental membrane conductance** is the product of the permeability of the membrane and the total membrane surface area.

Placental permeability and total membrane conductance changes during mammalian gestation. In early gestation, the placental membrane is thick and the placenta has not grown sufficiently. Since the permeability and the membrane conductance are low, the placenta is limited in the amount of substances it can diffuse between the fetal and maternal blood circulations. As gestation continues, there is a thinning of layers separating the fetal blood supply from the maternal blood supply, as well as an increase in the placental surface area for vasculature exchange. Therefore, there is an overall increase in the diffusion of nutrients and wastes between placental maternal and fetal blood vessels.

<u>Human Fetal Hemoglobin</u>

Three forms of hemoglobin (early embryonic, fetal and adult) appear during human fetal development (Figure 7.4). At approximately the 8th week of gestation, **early embryonic hemoglobin** is composed of a pair of zeta polypeptide chains and a pair of epsilon polypeptide chains. **Fetal hemoglobin** is composed of one pair of gamma polypeptide chains and one pair of alpha polypeptide chains. Fetal hemoglobin is produced from the 8th week of gestation to a time just before parturition. The production of **adult hemoglobin,** consisting of a pair of alpha polypeptide chains and a pair of beta polypeptide chains, is initiated just prior to parturition. By the 26th week of postnatal life, the blood of the infant consists entirely of adult hemoglobin.

<u>Gas Diffusion across the Placenta</u>

The amount of a gas that can diffuse across a permeable membrane per a given unit of time is best described using **Fick's Law for Diffusion** [Adolph Fick, German physician, 1829 – 1901]:

$$Vgas = \frac{(A)(D)(P1 - P2)}{T}$$

D = solubility of the gas/molecular weight of the gas

Vgas = amount of gas diffused per unit of time (ml/min)
A = area of the membrane available for diffusion
D = diffusion constant of a gas in the membrane
(P1 – P2) = partial pressure difference of the gas across the membrane
T = thickness of a membrane

In the human placenta, the partial pressure of oxygen in the maternal blood is 50 mm Hg, whereas the partial pressure of oxygen in the fetal blood is 20 mm Hg. Oxygen diffuses down the partial pressure gradient from the maternal blood into the fetal blood.

Although the partial pressure of oxygen is low in the fetal blood of the placenta, three physiological factors

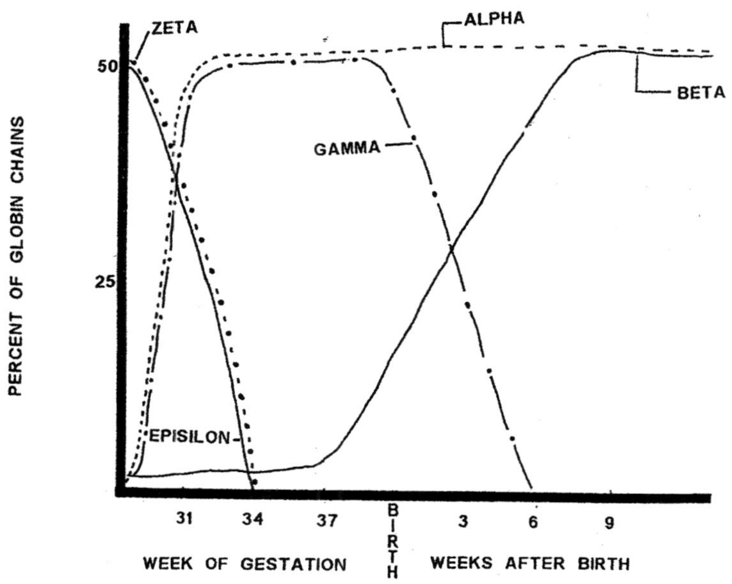

Figure 7.4. The development of the globin chains of human hemoglobin during gestation and early infancy.

enable the fetal blood to carry an adequate amount of oxygen to the fetal tissues. First, fetal hemoglobin shifts the oxygen dissociation curve to the left when compared to that of the adult (Figure 7.5). This means that at a given partial pressure of oxygen, fetal hemoglobin carries 20% to 30% more oxygen than adult hemoglobin. Secondly, the hemoglobin concentration of fetal blood is approximately 50% greater than that observed for the adult. Lastly, fetal blood contains large quantities of carbon dioxide. Diffusion of large amounts of carbon dioxide across the placenta from the fetal blood to the maternal blood makes the fetal blood of the placenta more alkaline and the maternal blood of the placenta more acidic. Since hemoglobin saturation in the blood is inversely related to blood pH, the decrease in maternal blood pH promotes more unloading of oxygen from hemoglobin, thereby causing more oxygen to diffuse across the placenta to the fetus.

The human fetus rids its body of carbon dioxide across the placenta. The partial pressure of carbon dioxide in the fetal blood is 2 mm Hg to 3 mm Hg higher than that of the maternal blood. Because of the high solubility of carbon dioxide in blood, a small partial pressure gradient is sufficient to allow large amounts of carbon dioxide to diffuse across the placenta from the fetal blood to the maternal blood.

<u>Movement of Nutrients and Wastes across the Placenta</u>

Glucose is transported across the human placenta from mother to fetus by the process of facilitated diffusion. In late pregnancy, the fetus utilizes more glucose than does the entire body of the mother. This causes the fetal blood glucose concentration to be 20% to 30% lower than that found in the maternal blood creating conditions for facilitated diffusion. The carrier molecules for glucose diffusion are found in the trophoblast cell membranes.

MAMMALIAN FETAL DEVELOPMENT

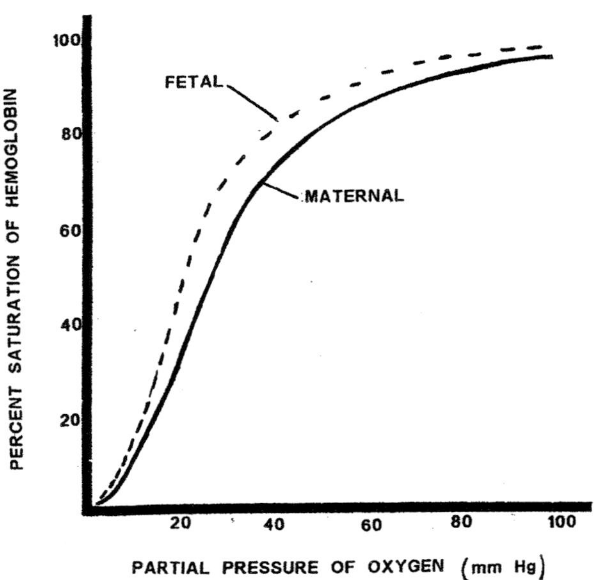

Figure 7.5. Oxygen dissociation curve for human maternal and fetal hemoglobins.

Fatty acids also diffuse across the placenta from the maternal blood to the fetal blood. The rate of fatty acid diffusion across the placenta is much slower compared to that observed for glucose. This is one reason glucose is the preferred energy source for the fetus.

The fetus eliminates its metabolic waste products across the placenta primarily by passive diffusion. Urea, uric acid and creatine diffuse from the fetal blood to the maternal blood. Because urea diffuses across the placenta with ease, the concentration of urea in the fetal blood is only slightly higher than that of in the maternal blood. Creatine does not diffuse across the placenta as readily as urea. This is the major reason why the level of creatine in the fetal blood is considerably higher than that found in the maternal blood.

MATERNAL AND PLACENTAL HORMONES DURING FETAL DEVELOPMENT

Introduction

In many mammalian species, the placenta eventually takes over the endocrine functions that was the responsibility of the corpus luteum. The human placenta produces large quantities of human chorionic gonadotropin, estrogen, progesterone, and human chorionic somatomammotropin. These hormones are all essential for pregnancy.

Human Chorionic Gonadotropin

Human chorionic gonadotropin (hCG) is a glycoprotein hormone that is secreted by the pituitary of the male and nonpregnant female in a pulsitile fashion similar to that observed for LH. Plasma levels of hCG are so low in the male and nonpregnant female that is difficult to detect in conventional assays. Human chorionic gonadotropin is secreted in the female during gestation. The trophoblastic cells of the human placenta secrete hCG into the body fluids of the mother. Human chorionic gonadotropin can be detected in the blood as early as 8 days to 9 days post ovulation. The concentration of human chorionic gonadotropin in the blood reaches a maximum by the 10th to 12th week of gestation, but decreases too much lower levels by week 16 (Figure 7.6). During the period of maximum blood concentration, hCG can also be detected in the urine.

Human chorionic gonadotropin prevents the normal involution of the corpus luteum. This hormone causes the corpus luteum to increase in size and to secrete larger amounts of estrogen and progesterone. Elevated plasma levels of estrogen and progesterone maintain the decidual nature of the uterine endometrium needed for early development of the conceptus. If the primate corpus luteum is removed before the 7th week of gestation, spontaneous abortion occurs. After that time, the placenta produces sufficient quantities of estrogen and progesterone to maintain the pregnancy for the remainder of the gestation.

Human chorionic gonadotropin influences the development of the male reproductive system during gestation. This hormone stimulates the interstitial cells of the testes to produce testosterone. Testosterone secreted during gestation is involved in the development of the male sex organs and is responsible for testicular descent.

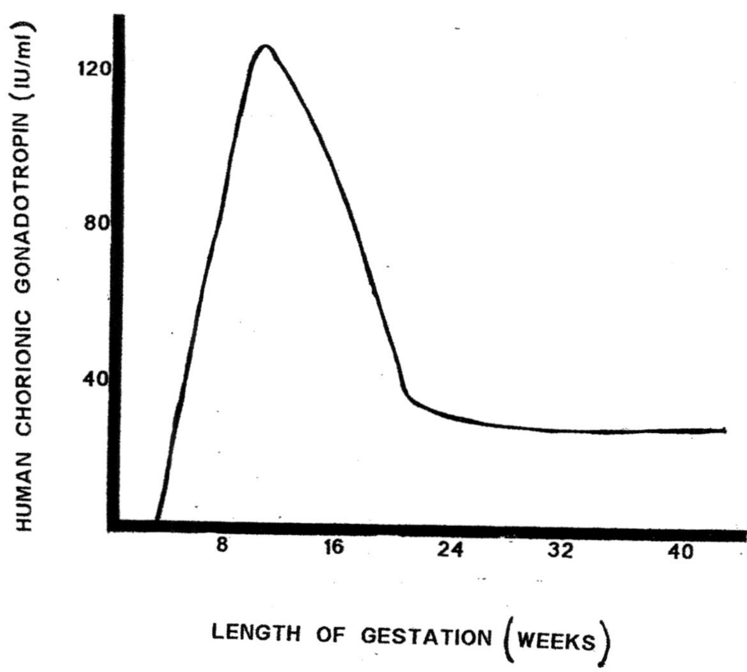

Figure 7.6. The concentration of hCG in maternal blood during human gestation.

MAMMALIAN FETAL DEVELOPMENT

Placental Estrogens

The trophoblastic cells of the placenta produce estrogens. The plasma concentrations of estrogens continue to increase during gestation until the time of parturition. (Figure 7.7). Secretion of estrogen by the placenta differs from that of the ovary in several ways. First, a large share of the placental estrogen secreted is estriol. Estriol concentration during gestation increases by a factor of a thousand compared to that found in the nonpregnant female. Estriol can be detected in the blood by the 9th week of gestation, reaching a plateau concentration from week 31 to week 35. At the 36th week of gestation, estriol concentration in the blood once again starts to increase until the end of gestation. Estriol is considerably less potent than estradiol–17B. During gestation, estrone increases 100 times over concentrations found in the nongravid female. Placental estrone can be detected in the blood as early as the 6th week of gestation. Estrone concentration then continues to increase throughout the remainder of gestation (Figure 7.7). During human gestation, plasma estradiol 17-B is approximately 100 times greater than that found in the nongravid female. Placental estradiol can be detected as early as the 6th week to the 8th week of gestation and continues to increase throughout the remainder of pregnancy (Figure 7.7).

Placental estrogens are not synthesized *de novo*. The production of placental estrogens is under the control of the fetus. (Figure 7.8.) The placenta lacks the 17–hydroxylation, 17, 20 desmolase activity as well as 16 alpha hydroxylase activity. This lack of enzymatic activity prevents the transformation of C-21 steroids into C-19 steroids. Androgenic compound dehydroepiandrosterone and 16-hydroxydihydroepiandrosterone produced by the fetal adrenal cortex and also of maternal origin are transported to the placenta and converted into estrogens (primarily estriol).

Estrogens have a proliferative effect during gestation. Estrogen causes enlargement of the uterus, breasts and external genitalia. Estrogen relaxes the various pelvic ligaments and sacroiliac joints become relatively limber. Estrogen causes the pubic symphysis to become elastic. Estrogen increases the rate of cellular mitosis during fetal development.

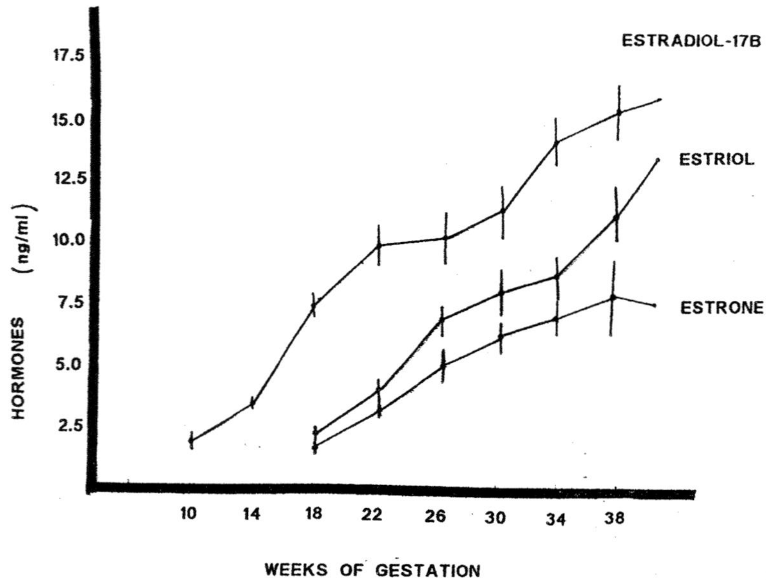

Figure 7.7. The concentration of maternal plasma estrogens during human gestation.

MAMMALIAN FETAL DEVELOPMENT

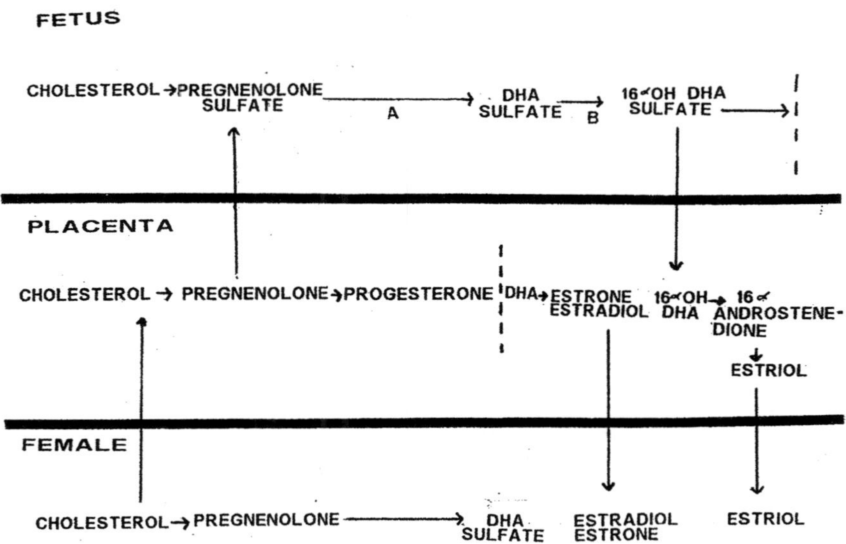

Figure 7.8. The production of estrogens during human gestation. Enzyme block (⋮), DHA (dehydroepiandrosterone).

Placental Progesterone

The corpus luteum and the placenta produce **progesterone** during human gestation. Progesterone secretion increases 10 fold during the course of gestation. The corpus luteum is responsible for moderate progesterone secretion from the moment of conception until approximately the 10th week of gestation. Human pregnancy is dependent on progesterone produced by the corpus luteum until the 7th week of gestation. Removal or degeneration of the corpus luteum prior to the 7th week of gestation results in the termination of pregnancy. Both the corpus luteum and the placenta from the 7th week of gestation to the 10th week of gestation share the production of progesterone. From the 10th week of gestation until the end of pregnancy, the placenta is the primary source of progesterone. The placenta is responsible for the secretion of high levels of progesterone during most of the gestation period (Figure 7.9).

The placental production of progesterone is independent of the fetus. Progesterone is formed from products found in maternal circulation (Figure 7.10). Approximately 97% of placental progesterone is synthesized from maternal cholesterol, whereas maternal pregnenolone is used to produce the remaining 3% of placental progesterone.

Progesterone is essential for the normal progression of pregnancy. Progesterone is involved in the development of the decidual cells of the uterine endometrium. Progesterone decreases the contractility of the gravid uterus helping to prevent spontaneous abortion. Progesterone contributes to the development of the conceptus prior to implantation by increasing the amount of nutritive substances secreted by the epithelium of the oviducts and uterus. Progesterone also prepares the mammary glands for lactation.

MAMMALIAN FETAL DEVELOPMENT

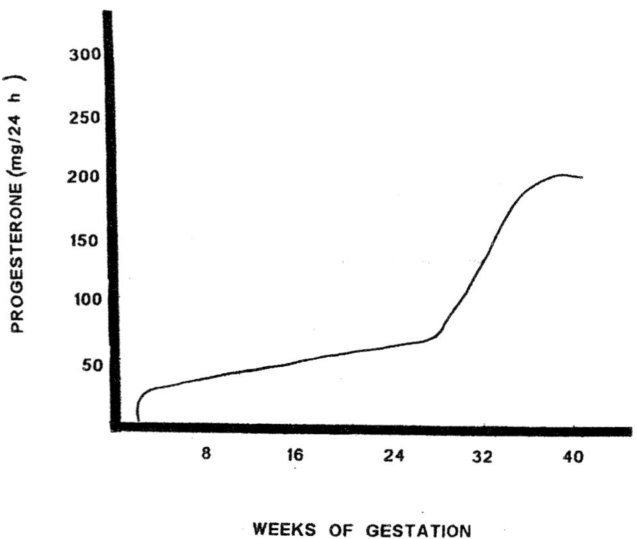

Figure 7.9. The concentration of maternal plasma progesterone during human gestation.

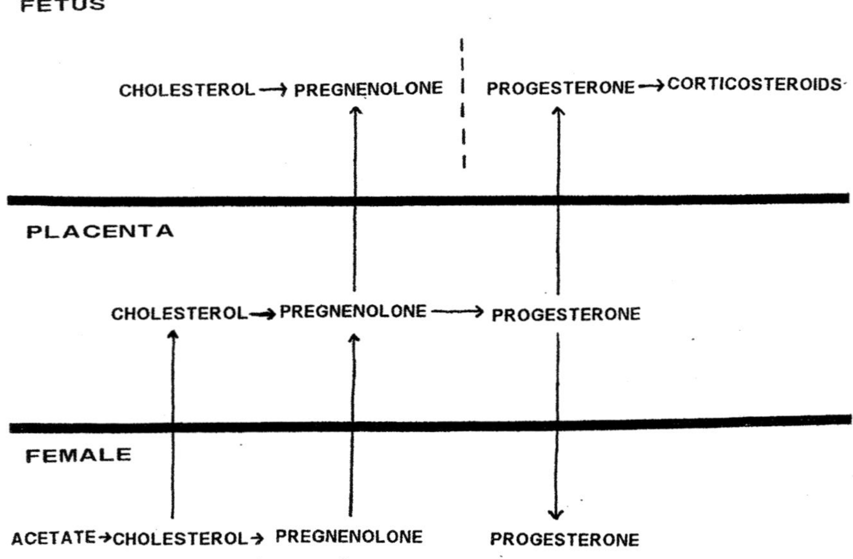

Figure 7.10. The synthesis of placental progesterone during human gestation. Enzyme block (┊).

116

MAMMALIAN FETAL DEVELOPMENT

<u>Human Chorionic Somatomammotropin</u>

Human chorionic somatomammotropin (hCS) is secreted by the syncytiotrophoblast of the placenta. Human chorionic somatomammotropin consists of 191 amino acids with 2 disulfide bonds. Human chorionic somatomammotropin has a molecular weight of approximately 32,000 and a half-life of approximately 15 minutes. The placenta begins to secrete hCS during the 5th week of gestation. The secretion of hCS is in direct proportion to the mass of the placenta and plateaus during the last four weeks of gestation (Figure 7.11).

Human chorionic somatomammotropin appears to be partially responsible for the development of the mammary glands, and in some species can initiate lactation. Because this function is similar to that observed for prolactin, hCS was once called **placental lactogen**. Attempts to promote lactation by administering hCS to humans have been unsuccessful.

Human chorionic somatomammotropin causes protein deposition in the uterus similar to that observed for growth hormone. Although its molecular structure is similar to that of growth hormone, hCS is approximately 100 times less sensitive than growth hormone in promoting growth.

During periods of maternal fasting, hCS is responsible for maintaining an energy flow to the developing fetus (Figure 7.12). Human chorionic somatomammotropin decreases maternal insulin directed entry of glucose into cells. This decrease in maternal sensitivity to insulin increases the amount of glucose available to the fetus. Some women may develop gestational diabetes that is most likely to occur during the third trimester when there is an increase demand for glucose by the fetus. After birth, blood glucose levels in human mothers may normalize. In other instances, blood glucose concentrations may remain impaired resulting in diabetes mellitus. Human chorionic somatomammotropin is also responsible for stimulating maternal lipolysis resulting in an increase in maternal levels of free fatty acids and ketones. Ketones are then transported across the placenta and used for energy by the developing fetus.

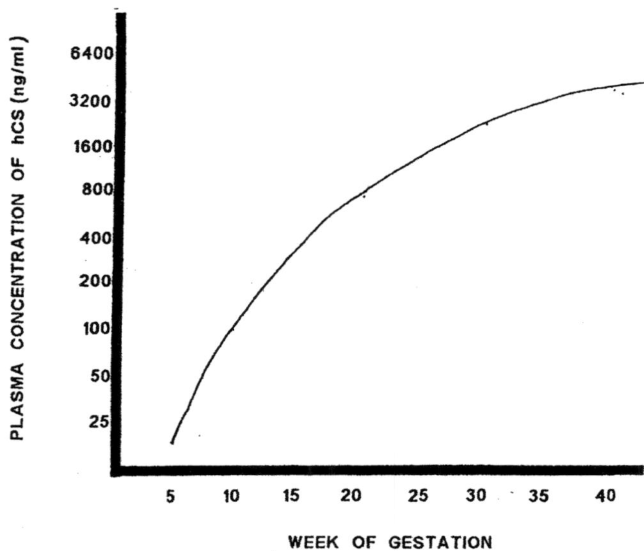

Figure 7.11. Maternal plasma concentration of hCS during human gestation.

A

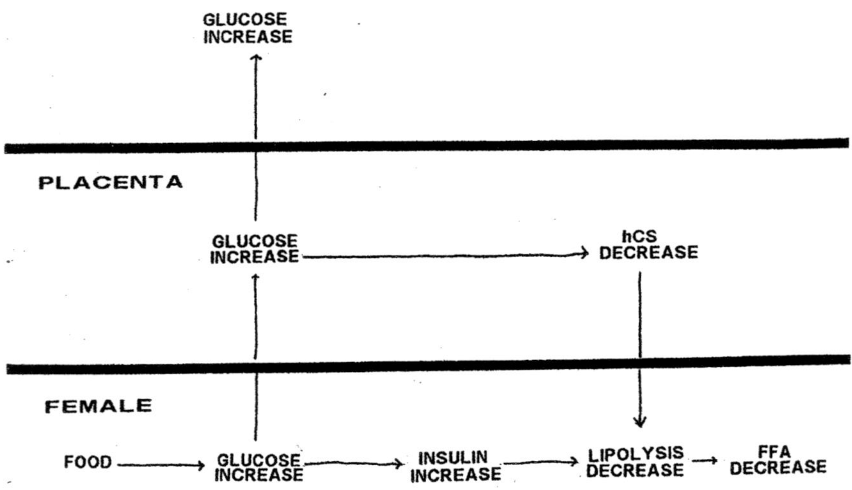

FFA = FREE FATTY ACIDS

B

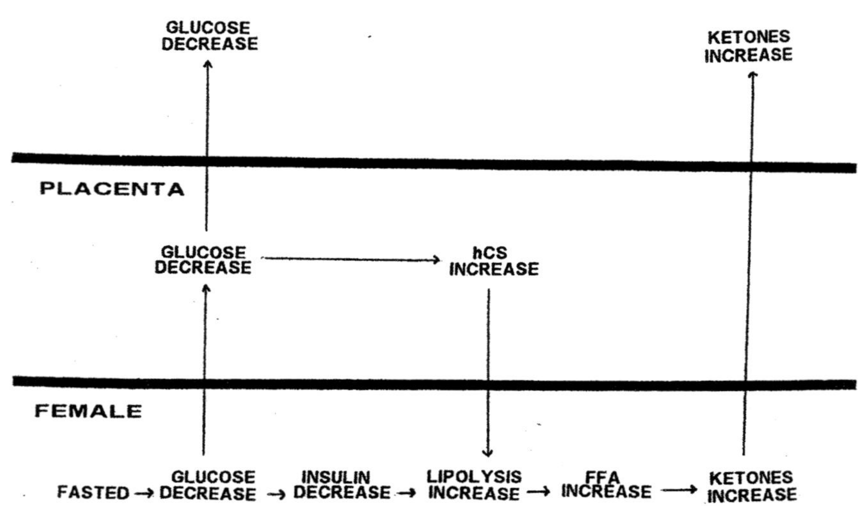

Figure 7.12. Regulation of placental hCS synthesis. (A) Maternal fed state. (B) Maternal fasted state.

MAMMALIAN FETAL DEVELOPMENT

<u>Placental Gonadotropin Releasing Hormone and Corticotropin Releasing Hormone</u>

Placental gonadotropic releasing hormone (placental GnRH) is a glycoprotein hormone found in the cytotrophoblast and syncytiotrophoblast of the human placenta. Placental GnRH stimulates the placental production of hCG. Placental GnRH also regulates the release of placental steroids and prostaglandins.

Placental corticotrophin releasing hormone (placental CRH) is produced by the trophoblast of the placenta, fetal membranes and decidua. The concentration of placental CRH in amnionic fluid and maternal circulation increases as human pregnancy progresses. The concentration of CRH binding protein, also produced by the trophoblast, fetal membranes and decidua, decreases in amnionic fluid and maternal circulation just prior to the onset of parturition thus increasing CRH activity. Placental CRH has been shown to increase the release of prostaglandin in the fetal membranes, decidua and the myometrium.

<u>Placental Growth Hormone, Growth Hormone Releasing Hormone and Somatostatin</u>

Growth hormone releasing hormone and somatostatin are found in the human placenta during gestation. Somatostatin is also found in the decidua. The human placenta also produces growth hormone. Placental growth hormone replaces maternal growth hormone by term. Changes in the level of maternal IGF-1 and IGF-binding proteins reflect changes in placental growth hormone regulation. Placental growth hormone synthesis is regulated by maternal glucose levels and not by placental GHRH.

<u>Maternal Secretion of Glucocorticoids and Aldosterone during Gestation</u>

Maternal adrenal secretions also play a role during gestation. Glucocorticoid secretion increases moderately throughout pregnancy. It is believed that these hormones help mobilize amino acids from the maternal tissues so they can be utilized in fetal metabolism. There is an increase in the maternal plasma levels of aldosterone throughout pregnancy. Aldosterone causes the kidneys of pregnant women to reabsorb excess sodium thereby causing fluid retention. This occasionally leads to maternal systemic hypertension.

<u>Maternal Secretion of Parathyroid Hormone and Relaxin during Gestation</u>

The developing human fetus removes calcium from the female to build its own skeleton. This becomes particularly pronounced during the last third of gestation. The maternal parathyroid gland enlarges (especially in calcium deficient women) during the course of gestation. Increased maternal parathyroid hormone secretion causes calcium absorption from maternal bone tissue helping the female maintain normal extracellular calcium levels as the fetus removes calcium from her body. The secretion of parathyroid hormone is more intense during lactation since the newborn requires more calcium than the developing fetus.

Relaxin is a polypeptide with a molecular weight around 9000. The corpus luteum and the placenta secrete relaxin. In the rat, relaxin causes relaxation of the ligaments of the pubic symphysis helping to facilitate parturition. In human females, the role of relaxin during gestation and parturition is poorly understood. Relaxin has been shown to be responsible for softening of the human cervix during late gestation.

MAMMALIAN FETAL DEVELOPMENT

DEVELOPMENT OF THE FETAL NERVOUS SYSTEM

<u>Neural Cell Generation, Migration and Formation of Specific Neuronal Populations</u>

The neuroepithelium of the neural tube is composed of one or two layers of columnar epithelial cells. The cells of the neuroepithelium mitotically divide at the time of closure of the neural tube. Most of the cellular proliferation occurs in the ventricular zone (the inner surface of the neural tube) with elongated cellular processes that expand the entire wall of the neural tube. The nuclei of neurons migrate towards the marginal zone, then replicate, before migrating back to the ventricular zone. The cytoplasmic process of each neuron retracts prior to producing two daughter cells.

After several proliferative cycles, precursor neuronal cells lose their ability to divide and migrate away from the germinal zone to occupy their position in the developing nervous system. In many parts of the central nervous system radial glia (special support cells) form a scaffold complex guiding the migration of immature neurons. In other parts of the central nervous system and peripheral nervous system immature neurons migrate to their appropriate positions in the developing nervous system using molecular signals in the extracellular matrix.

Proliferative cells of the neuronal tube produce an enormous array of neuron phenotypes during mammalian development. During human gestation there appears to be two distinct periods of cellular proliferation. The first period of cellular proliferation is the major period of neurogenesis occurring between the 10th to the 18th week of gestation. The second period of cellular proliferation is a period of gliogenesis that occurs from midgestation to approximately 24 months of age (glial precursors and radial glial cells are present in the human nervous system from early gestation). For neurons in the brain forming the cortex of the cerebrum, the hippocampus, and the optic tectum, the birthdate of the neuron will determine its laminar fate. The first neurons to develop will form the deepest layers of that particular structure, whereas those neurons that are generated later during gestation will form progressively more superficial layers. In the cerebellum, larger neurons such as the Purkinje cells [Johannes E. von Purkinje, Bohemian anatomist and physiologist, 1787 – 1869] develop before smaller neurons such as the cerebellar granule cells. In the spinal cord there is a ventral to dorsal progression of cell proliferation. The proliferation of ventrally located efferent neurons of the spinal cord is completed before that of afferent neurons in the dorsal horn.

<u>Development of Neural Processes</u>

Once young neurons have migrated to specific positions in the developing nervous system they must form connections with other targets. The **growth cone** is a structure at the end of a developing neuronal process (Figure 7.13). The growth cone is a **lamellipodium** from where **filopodia** extend. The filopodia of the growth cone are typically 5 micrometers to 30 micrometers in length with a diameter of approximately 0.1 micrometers. Fibrillar actin fills the filopodia and organelles fill the leading edge of the growth cone. Growth cones move by the extension and contraction of the filopodia. The mechanism by which the leading edge of the growth cone advances is unclear. A "tread-like" mechanism may possibly be involved in the advancement of the growth cone filopodia. This tread-like mechanism of filopodia advancement may involve actin monomers that polymerize at the leading edge of the growth cone and are then swept towards the rear of the growth cone where they are then depolymerized and recycled back to the leading edge of the growth cone for repolymerization.

Axonal pathfinding involves external short-range cues and external long-range cues that each may be either attractive or repulsive. Axons have receptors on the surface of the growth cone that can respond to these

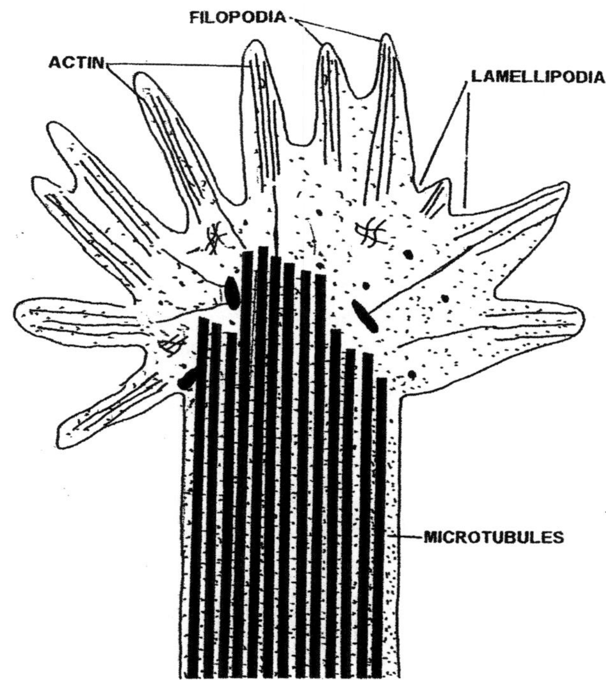

Figure 7.13. A neuron growth cone.

external cues. In short-range attraction, when a receptor molecule on a growth cone finds a substance in the external environment in which to bind, forward movement of the growth cone occurs by the local insertion of membrane components at the growth cone. In long-range attraction a variety of cells can release chemotropic factors that are diffusable and attract neuronal axons. In short-range repulsion, neurons can have growth cone receptors that bind to ligands on the surface of a second neuron. The binding of the receptor and the ligand results in the axon being repelled rather than attracted. In long-range repulsion, growth cones can be repelled by diffusable factors secreted from a distance.

<u>Synaptogenesis</u>

Synaptogenesis is the formation of a connection between two excitable cells. Synaptogenesis often depends on **cell specificity**. Axons generally migrate through the nervous system but do not form synapses with everything along their paths. Specific types of neurons are attracted to other similar neurons due to the presence of specific marker molecules. Synaptogenesis can also depend on **topographical specificity**. In some areas of the nervous system (e.g. eye and somatosensory cerebral cortex), axons of presynaptic neurons must connect to postsynaptic cells in a highly ordered manner to maintain positional information. The creation of this topographical specificity requires a gradient of signaling molecules be present to encode positional information. Synaptogenesis often requires **synaptic site specificity** where the axon of a presynaptic neuron must synapse on a specific region of the postsynaptic cell. Synaptic site specificity often depends on the arrival time of a particular afferent fiber on a particular region of the postsynaptic cell. Synaptic site specificity appears also be the result of afferent fibers competing for specific cellular and/or molecular markers on postsynaptic cells. In the nervous system of many eutherian mammals, synaptogenesis often peaks in the immediate postnatal period.

MAMMALIAN FETAL DEVELOPMENT

To create a synapse between two excitable cells, one or more filopodia of the presynaptic cell makes contract with the postsynaptic cell. There is a cessation of all filopodial activity of the presynaptic cell for approximately 30 minutes. In the region of the growth cone of the presynaptic axon, presynaptic dense projections develop and synaptic vesicles are present by approximately 18 hours post contact. Postsynaptic specializations begin to develop within 6 hours post contact.

<u>Neuronal Regression during Fetal Development</u>

Development and shaping of the nervous system in eutherian mammals requires neuronal death during gestation and early postnatal life. The death of neurons appears to occur in response to the production of various cytoplasmic proteases and endonucleases that destroy the DNA. Apoptosis (programmed cell death) has been observed in many areas of the nervous system and often occurs during a specific time period for specific neural populations. The purpose of neuronal apoptosis can be several. First, the major purpose appears to be matching the size of the innervating population of neurons to the capacity of their targets. Secondly, apoptosis corrects for errors of mislocation and misprojection.

Neuron apoptosis in the developing nervous system can be target dependent or hormone dependant. In **target dependent neuronal apoptosis**, target cells appear to produce limited amounts of sustaining substances (e.g. nerve growth factor and ciliary neurotropic factor). Neurons appear to compete for these limited amounts of sustaining substances and only small portions of the total neurons in that area survive. Trophic neuronal support can also be derived from autocrine or paracrine sources of secretions. In **hormone dependent neuronal apoptosis**, death of neurons is directly or indirectly related to the action of circulating hormones. In some parts of the nervous system related to reproduction (e.g. spinal nucleus of the bulbocavernosus) loss of efferent neurons is greater in the female compared to that of the male due to the lack of androgen hormones.

After neuronal death there is usually an exuberant growth of synaptic connections between surviving cells and the selective elimination of neuronal processes without death of neurons. This elimination of axons and their synapses occur over relatively short distances and usually occur in the early postnatal period in most large mammalian species.

DEVELOPMENT OF THE FETAL ENDOCRINE SYSTEM

<u>Development of the Fetal Hypothalamus and Adenohypophysis</u>

The human fetal hypothalamus produces releasing and inhibiting hormones that are transported by the hypothalamic-hypophyseal portal system to influence the secretion of hormones by the fetal adenohypophysis. The human fetal hypothalamic-hypophyseal portal system is present by approximately midgestation. The production of fetal ACTH by corticotrophs of the fetal adenohypophysis is stimulated by CRH and arginine vasopressin released from the parvocellular cells of the paraventricular nuclei of the fetal hypothalamus. The medial preoptic area of the fetal hypothalamus produces GnRH that stimulates the gonadotrophs of the fetal adenohypophysis to produce and release FSH and LH. The paraventricular nuclei of the fetal hypothalamus produce TRH that stimulates the thyrotrophs of the fetal adenohypophysis to secrete and produced TSH. The arcuate nuclei of the fetal hypothalamus produce GHRH that stimulates the somatotrophs of the fetal anterior pituitary to secrete GH. The paraventricular nuclei of the fetal hypothalamus produce somatostatin that inhibits the secretion of GH in the fetal adenohypophysis. The arcuate nuclei of the fetal hypothalamus secrete dopamine that tonically inhibits the secretion and synthesis of prolactin by the lactotrophs of the fetal adenohypophysis.

MAMMALIAN FETAL DEVELOPMENT

Development of the Fetal Neurohypophysis

The supraoptic nucleus and the magnocellular region of the paraventricular nucleus of the human fetal hypothalamus secrete arginine vasopressin and oxytocin. Oxytocin and arginine vasopressin is transported to the fetal neurohypophysis via the hypothalamic-hypophyseal tract.

Fetal oxytocin circulates in the plasma of the fetus in a processed and unprocessed form. As gestation proceeds, the proportion of processed oxytocin increases in the fetal plasma. Fetal oxytocin is not permeable across the placenta, thus fetal and maternal plasma concentrations of oxytocin are regulated independently. The function of fetal oxytocin is unclear. Since fetal oxytocin does not cross the placenta, it appears to have no influence on uterine muscle tone. High concentrations of fetal oxytocin stimulate the release of ACTH from fetal corticotrophs. Fetal oxytocin may serve as a releasing factor for fetal ACTH.

Arginine vasopressin (AVP) is a hormone with three biological functions. Arginine vasopressin has a vasopressor effect on the peripheral vasculature. Arginine vasopressin can also stimulate the release of ACTH from corticotrophs of the adenohypophysis. Arginine vasopressin also has an antidiuretic effect on the kidney. Each of these biological effects occurs in response to stimulation of different receptors that bind AVP. Human and ovine fetuses excrete large volumes of dilute urine suggesting that AVP is less potent on the kidney of the fetus compared to that of the adult. Arginine vasopressin is also important in fetal cardiovascular regulation. During periods of fetal hemorrhaging or systemic hypotension, plasma levels of fetal AVP increase. Increased levels of fetal AVP appear to promote the redistribution of fetal combined ventricular output towards the umbilical-placental circulation thus maintaining the diffusion of oxygen from maternal blood to fetal blood.

Development of the Fetal Hypothalamic-Hypophyseal -Thyroid Axis

In many mammalian species that the fetal hypothalamic-hypophyseal-thyroid axis is functional by midgestation. During fetal development in sheep, development of thyrotrophs occurs in the first third of gestation. The thyroid gland of the fetal lamb is developed when gestation is approximately 40% complete and the hypothalamus is secreting TSH by midgestation. In humans, thyrotropin (TSH) is present in the fetal adenohypophysis and blood plasma by midgestation. The maternal hypothalamic-pituitary-thyroid axis and the fetal hypothalamic-pituitary-thyroid axis operate somewhat independently. Tetraiodothyronine (T_4) and triiodothyronine (T_3) are relatively impermeable across the placenta. Thyroid binding globulins and TSH are also relatively impermeable across the placenta. During fetal development the secretion of T_4 is under the influence of fetal TSH. The level of thyroxin increases from the 20th week of gestation to levels exceeding maternal concentrations at term. From the second trimester until birth, the levels of maternal thyroxin do not meet the demands of the fetus. Fetal thyroxin is needed for normal development of the many body systems. Low levels of fetal thyroxin results in poor bone development, lack of body hair and impaired differentiation of the central nervous system. Fetal thyroxin is important in preparing the fetus for extrauterine life. Removal of the thyroid gland from fetal lambs results in severe hypothermia after birth. Severe hypothermia occurs in response to impaired thermogenesis in brown tissue. Fetal calcitonin is in relatively high concentrations in the fetal blood plasma during gestation and declines shortly after birth. During gestation, fetal calcitonin is important in preventing hypocalcemia. Abnormally elevated plasma concentrations of fetal calcitonin causes severe fetal hypocalcemia.

Development of the Fetal Hypothalamic-Hypophyseal-Adrenal Axis

The human fetal hypothalamic-hypophyseal –adrenal axis is similar to that found in the adult. Arginine vasopressin and CRH are released from the paraventricular nuclei of the fetal hypothalamus and are transported to the adenohypophysis via the hypothalamic-hypophyseal portal system. Fetal AVP and CRH then stimulate fetal corticotrophs of the adenohypophysis to secrete ACTH. In humans, AVP and CRH increase the sensitivity of the corticotrophs to the other hypothalamic-releasing hormone, respectively. Adrenocorticotropic hormone is then transported in the blood to the fetal adrenal gland. Fetal ACTH has an inductive role in the growth and development of the fetal adrenal cortex. Fetal ACTH stimulates the production and release of mineralocorticoid and glucocorticoid hormones from the fetal adrenal cortex.

Adrenocorticotropic hormone is a protein that does not cross the placenta. Therefore, maternal plasma ACTH is of maternal origin and fetal plasma ACTH is of fetal origin. However, maternal cortisol does cross the placenta to some extent. This allows maternal cortisol to maintain fetal homeostasis during early gestation until the fetal hypothalamic-hypophyseal –adrenal axis is mature enough to maintain its own levels of cortisol.

Human **fetal aldosterone** becomes elevated during late gestation but does not appear to become responsive to changes in blood volume until after birth. Human **fetal cortisol** can be found in the fetal blood as early as the 10th week of gestation and gradually increases during gestation reaching a peak concentration during parturition. Human **fetal glucocorticoids** have numerous functions during gestation. Fetal glucocorticoids are involved in the induction of enzymes needed for surfactant synthesis in the lungs. Fetal glucocorticoids are involved in the stimulation of respiratory alveolar water absorption as well as the stimulation of central respiratory mechanisms. Fetal glucocorticoids are involved the induction of enzyme systems involved in gluconeogenesis and glucose storage in the liver and the myocardium. Fetal glucocorticoids are involved in the maturation of the pancreatic islets and the secretion of insulin by beta cells. Fetal glucocorticoids are involved in the development of the lobules, ducts and alveoli of the mammary gland. Fetal glucocorticoids are involved in the synthesis of epinephrine by the induction of phenylethanolamine-N-methyl transferase in the adrenal medulla. The conversion of tetraiodothyronine to triiodothyronine in the fetus appears to be influenced by the fetal glucocorticoids. The switch in the production of fetal hemoglobin to the adult hemoglobin appears to occur under the influence of fetal glucocorticoids. Fetal glucocorticoids are also responsible for the onset of hemopoiesis in the bone marrow. Fetal glucocorticoids are involved in the parturition process that will be discussed later.

Development of the Fetal Adrenal Medulla

The development of the human fetal adrenal medulla occurs along with other components of the sympathetic nervous system. The fetal adrenal medulla secretes epinephrine and norepinephrine. The secretion of epinephrine from the fetal adrenal medulla increases in late gestation in response to the activation of phenylethylamine-N- methyltransferase. Activation of phenylethylamine –N-methyltransferase is dependent on the secretion of cortisol from the fetal adrenal cortex.

Fetal adrenal secretion of epinephrine and norepinephrine is important in helping to maintain homeostasis during periods of fetal stress. In response to fetal hypoxia, sympathoadrenal activity aids in vasoconstriction that redistributes the combined ventricular output of the fetus towards the umbilical-placental circulation. An increase in the secretion of epinephrine from the fetal adrenal medulla causes the liver to release glucose into the fetal blood thus increasing the plasma concentration of glucose.

Development of the Fetal Hypothalamic-Hypophyseal-Gonadal Axis

The anatomy and physiology of the human fetal hypothalamic-hypophyseal-gonadal axis is similar to that found in the adult. Gonadotropic releasing hormone is released from the median eminence of the hypothalamus into the hypothalamic–hypophyseal portal system and transported to the adenohypophysis. Gonadotropic releasing hormone stimulates the gonadotrophs to release FSH and LH. Plasma concentrations of the gonadotropic hormones increase reaching peak concentrations near midgestation then decreasing until the time of parturition (Figure 7.14). The concentration of the gonadotropic hormones during fetal development is higher in the female when compared to the male. The lower concentration of plasma gonadotropic hormones in the male fetus occurs in response to the negative feedback effects of testosterone secreted by the testes.

Circulating androgens and estrogens in the mammalian fetus are of gonadal and placental origin. The gonadal production of sex steroids in the fetus peaks at approximately midgestation and is responsible for gonadal growth and differentiation. The hypothalamic-hypophyseal-adrenal axis and the placenta control the production of androgens and estrogens in later gestation.

Development of the Fetal Parathyroid Gland

The parathyroid gland of the human fetus begins to secrete parathyroid hormone by the 12th week of gestation, but plasma concentrations remain relatively low during gestation. The suppression of the fetal parathyroid hormone occurs in response to high circulating levels of maternally derived calcium.

Development of the Fetal Endocrine Pancreas

The human fetal pancreas is active early in gestation. Prior to the 10th week of gestation, alpha cells and gamma cells of the pancreatic islets are secreting glucagon and somatostatin, respectively. Beta cells that produce insulin can be observed by the 10th week of gestation. Initially each group of cells is clustered separately in the islets. Only later do the beta cells become surrounded by alpha and gamma cells. From

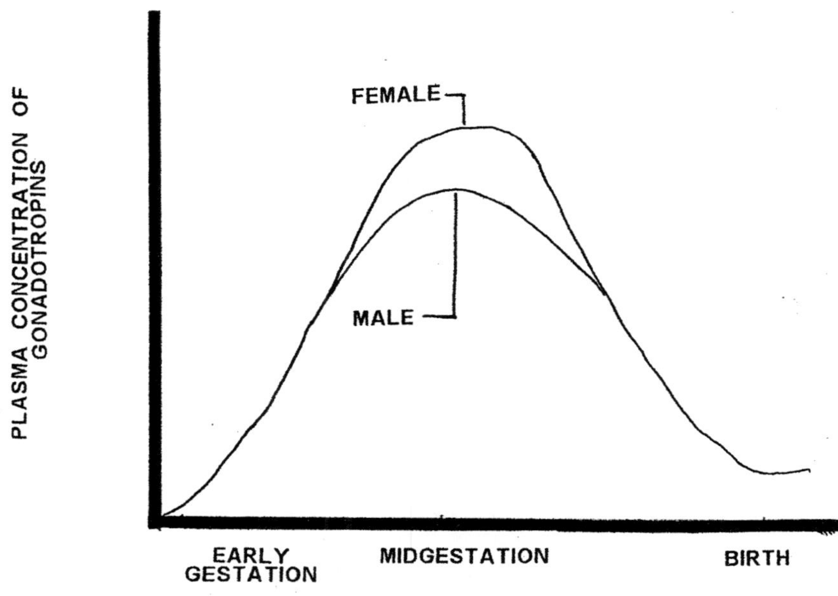

Figure 7.14. Plasma concentration of fetal gonadotropins during human gestation.

MAMMALIAN FETAL DEVELOPMENT

approximately midgestation until term, the secretion of insulin from the beta cells occurs in response to elevated plasma levels of amino acids, glucose and short chain fatty acids. Inhibition of insulin secretion occurs in response to catecholamines.

<u>Alpha Fetoprotein</u>

Human **alpha-fetoprotein** is a glycoprotein composed of 590 amino acids. Carbohydrates comprise approximately 4% of the molecule. The yolk sac and the fetal liver produce human alpha-fetoprotein. The concentration of alpha-fetoprotein in human fetal blood peaks at the end of the 1st trimester. From the end of the first trimester until week 32 of gestation, the concentration of human alpha-fetoprotein gradually decreases in the fetal blood. After the 32nd week of gestation, the concentration of alpha-fetoprotein rapidly declines in the blood of the fetus. The function of alpha-fetoprotein in the human fetus remains unknown. Alpha-fetoprotein may serve as a carrier of steroids in the fetal blood.

DEVELOPMENT OF THE FETAL CARDIOVASCULAR SYSTEM

<u>Development of the Fetal Cardiovascular System</u>

The human heart begins to function during the 4th week of gestation, contracting at a rate of approximately 65 beats per minute. The cardiac rate of the human fetus steadily increases reaching about 140 beats per minute prior to birth. Nucleated erythrocytes are being formed in the yolk sac by the 3rd week of gestation, followed a week later by the formation of non-nucleated erythrocytes by the fetal mesenchyme. By the 3rd month of gestation erythrocyte and leukocyte production can be observed in the bone marrow, liver and lymphoid tissues. During the 3rd trimester, the bone marrow gradually takes over red blood cell production, whereas many of the extra-marrow areas loose their ability to form erythrocytes.

<u>Anatomy of Fetal Circulation</u>

The circulatory system of the fetus differs from that of the adult (Figure 7.15). Because the placenta is involved in nutrient, gas and waste exchange between the fetus and the mother, blood flow is shunted way from the fetal lungs and liver during gestation. Blood returning from the placenta to the fetus through the umbilical vein is oxygenated and nutrient rich. Two separate paths carry oxygenated and nutrient rich blood from the placenta to the inferior vena cava. The **ductus venosus** bypasses the fetal liver and delivers most of the blood to the inferior vena cava. A small amount of blood from the placenta perfuses the liver and enters the inferior vena cava through the hepatic vein.

Blood from the inferior vena cava enters the right atrium of the fetal heart. The **cristae dividens** (free edge of the interatrial septum projecting from the foramen ovale) divides the blood flow of the right atrium into two streams. Most of the blood entering the heart from the inferior vena cava is transported to the left atrium through the **foramen ovale** (hole in the interatrial septum). Most of the blood entering the right atrium of the heart via the coronary sinus is also transported through the foramen ovale to the left atrium. A small amount of blood entering the right atrium of the heart via the inferior vena cava enters the right ventricle. Blood leaving the right ventricle of the fetal heart is divided into two streams. Most of the blood from the right ventricle is transported to the aorta via the **ductus arteriosus** (small duct transporting blood from the pulmonary trunk to the aorta). A small volume of blood from the right ventricle of the fetal heart is transported to the lungs by the pulmonary arteries.

MAMMALIAN FETAL DEVELOPMENT

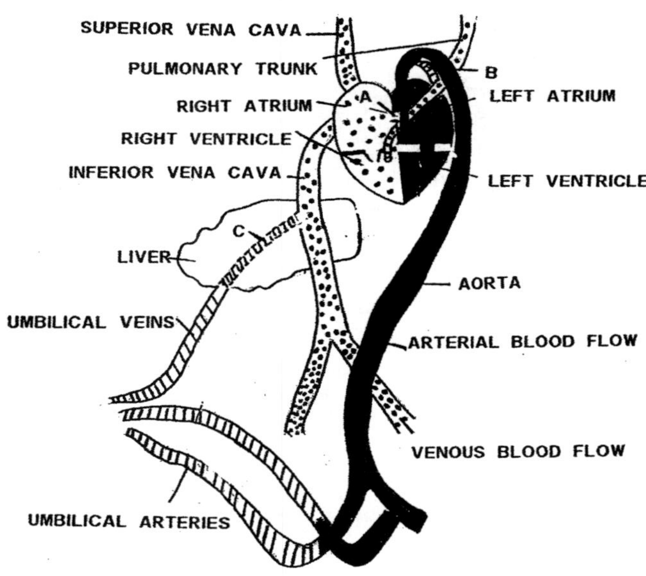

Figure 7.15. Circulatory system of the human fetus. (A) Foramen ovale. (B) Ductus arteriosus. (C) Ductus venosus.

The fetal circulatory shunts described above are responsible for the distribution of oxygenated blood to the head and body. Blood leaving the placenta via the umbilical veins is approximately 90% saturated with oxygen. A decrease in oxygen saturation (67%) is noted in blood entering the lower inferior vena cava. This decrease in blood oxygen saturation in the lower inferior vena cava is due to the mixing of umbilical blood with oxygen poor blood (approximately 20% saturated with oxygen) from the hepatic vein. Oxygen poor blood (approximately 31% oxygen saturation) from the head region of the body mixes with blood entering the right atrium of the fetal heart via the inferior vena cava. This mixing of blood in the right atrium lowers the oxygen saturation of the blood to approximately 52%. Blood entering the aorta via the ductus arteriosus mixes with oxygen rich blood in the aorta to yield blood in the descending aorta that is approximately 58% saturated with oxygen.

DEVELOPMENT OF THE FETAL RESPIRATORY SYSTEM

Anatomical Development of the Fetal Lung

Lung development in many mammals can be divided into stages based on microscopic development. Structural development of the lung is a continuous process that starts during embryonic development and continues well beyond birth in many mammalian species. These stages of lung development for three mammalian species are discussed below.

The **embryonic stage of lung development** occurs from conception to the 6th week of gestation in humans, from conception to day 13 in rats, and from conception until day 40 in the fetal lamb. Primitive lung buds appear as an outgrowth from the ventral surface of the embryonic foregut. The primitive lung bud rapidly divides to form a branched tube-like structure that eventually gives rise to the epithelial lining of the larynx, trachea, bronchioles and lower airways of the lung. The branching epithelial tubes invade the splanchnic mesoderm that eventually gives rise to the nonepithelial structures of the respiratory airways.

The **pseudoglandular stage of lung development** occurs from the 6th week of gestation to the 16th week of gestation in humans, from day 13 to day 18 of gestation in rats, and day 40 to day 80 of gestation in the fetal lamb. The pseudoglandular stage of lung development is characterized by a rapid proliferation of the airways at the distal end of the epithelial tubes. This progressive maturation of lung tissue begins at the lung hilus and radiates towards the periphery of the lung. The primitive airways are lined with tall columnar epithelium and the lumens are filled with liquid. The primitive epithelial cells lining the primitive airways start to differentiate into ciliated epithelial cells, goblet cells and mucous glands. The primitive alveoli starting to develop are surrounded by a poorly vascularized mesenchyme.

The **canalicular stage of lung development** occurs from the 16th week of gestation to the 26th week of gestation in humans, from day 18 of gestation to day 20 of gestation in rats, and from day 80 of gestation to day 120 of gestation in the fetal lamb. The canalicular stage of lung development is characterized by the increase in diameter and length of the respiratory airways. Further development of the terminal and respiratory bronchioles eventually occur forming terminal saccules. Thinning and increased vascularization of the mesenchyme occurs around the terminal alveoli. Epithelial differentiation can be observed in the terminal airsacs during this period of time. Cells that eventually form type I alveolar cells start to elongate forming thin cytoplasmic extensions that provide a thin barrier for external respiration. Cells that eventually form type II alveolar cells remain spherical in shape and develop lamellar bodies that contain **surfactant**. Towards the end of the canalicular stage of lung development external respiration is possible.

The **alveolar/saccular stage of lung development** occurs from the 26th week of gestation until after birth in humans, between day 20 of gestation until after birth in rats, and between day 120 of gestation until after birth in the fetal lamb. Terminal saccules progressively subdivide forming individual alveoli due to remodeling of the saccule walls. Extensive capillary growth ensures that each alveolus is well vascularized. Interstitial tissue between the alveoli and the pulmonary capillaries continues to thin, aiding in the formation of a thin barrier for external respiration.

<u>Breathing Movements of the Fetus</u>

Rhythmic contraction of the muscles (especially the diaphragm) that eventually will be involved in pulmonary ventilation can be observed as early as the 11th week of gestation in the human fetus. **Fetal breathing movements** (FBM) become discontinuous especially in late gestation. Fetal breathing movements primarily occur during periods of low voltage cerebral electrocortical activity and rapid eye movement. Fetal breathing movements become episodic after the 28th week of gestation in humans and after 120 days of gestation in the fetal lamb in response to long periods of apnea. High voltage cerebral electrocortical activity and non-rapid eye movement (REM) periods are believed to inhibit the respiratory centers of the brain. The amount of time the fetus spends making breathing movements varies between individuals. Factors that influence FBM include gestational age, time of the day, and maternal behaviors. Although FBM cause a slight decrease in intrapleural pressure, little liquid is inhaled partially due to the high viscosity of the amnionic fluid.

<u>The Production and Control of Fluid Secretion from the Fetal Lung</u>

Fluid fills the airways of the lungs from early gestation until the time of parturition. Fluid filling the respiratory airways of the fetal lung are a secretory product of the pulmonary epithelium. Fetal lung fluid leaves the lung, flows up the trachea into the pharynx. Fetal lung fluid is either swallowed or enters the amnionic sac.

At the level of the alveolus, the tight junctions occurring at the apical surface of pulmonary epithelial cells separate the lung lumen from the underlying interstitium. Sodium-potassium-ATPase pumps located on the basolateral surface of pulmonary epithelial cells generate an ionic sodium gradient from the exterior to the interior of the cell (Figure 7.16). Ionic sodium linked with ionic chlorine and ionic potassium enter the pulmonary epithelial cells. Ionic chlorine moves out of the pulmonary epithelial cells down its electrochemical gradient through channels located on the apical surface. The movement of ionic chlorine into the lumen of the alveolus creates an osmotic gradient for movement of fluid into the lumen of the alveolus.

The state of fetal activity influences the secretion of fluid from the fetal lung. During periods of apnea, the larynx is constricted creating a high resistance to the efflux of fluid from the fetal lung. This promotes the accumulation of lung fluid in the lower airways of the lung. During periods of fetal breathing movements, the larynx dilates reducing resistance to the efflux of fluid from the fetal lung. This allows lung fluid to exit the lung at an increased rate.

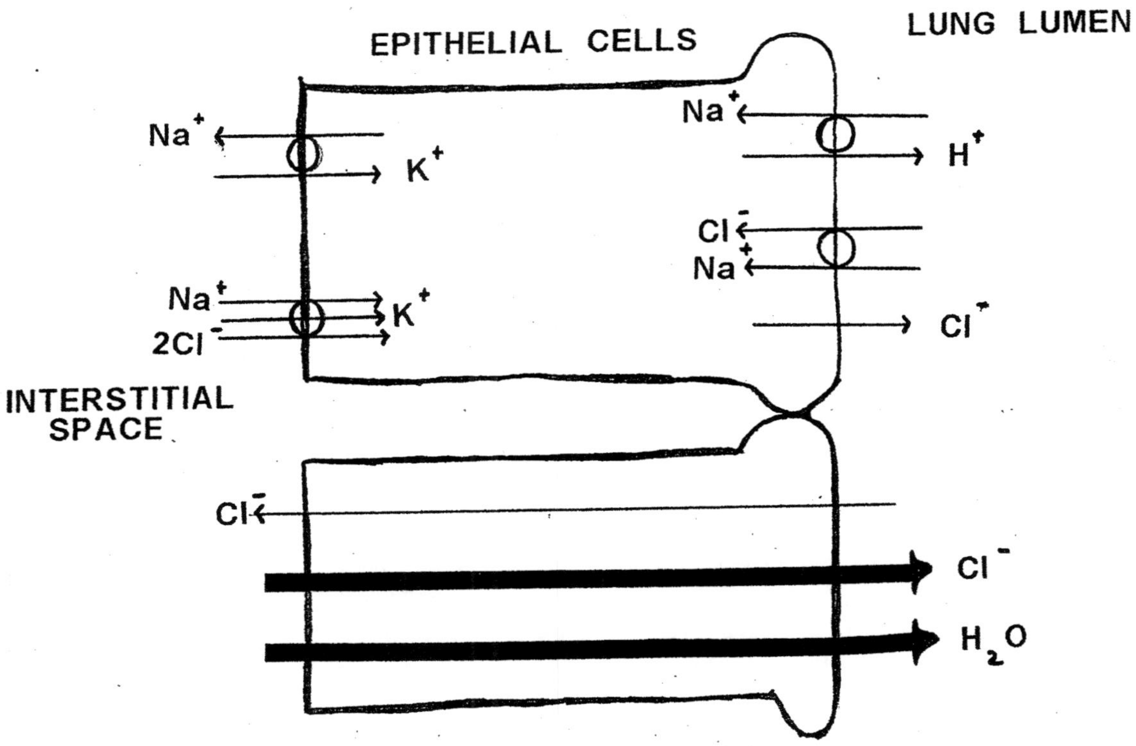

Figure 7.16. Fluid secretion in the human fetal lung.

MAMMALIAN FETAL DEVELOPMENT

<u>The Role of Lung Liquid in the Growth of the Fetal Lung</u>

Expansion of the fetal lung is important in stimulating growth. Tracheal occlusions in fetal lambs cause fluid to accumulate in the airways of the lung resulting in expansion of the lung. Lung expansion is a stimulus for cellular replication. The increase in cellular replication parallels the increase in lung liquid volume. Draining of liquid from the fetal lung essentially causes lung growth to cease.

<u>The Role of Surfactant in the Fetal Lung</u>

Surfactant is a complex mixture of phospholipids and proteins and is responsible for reducing surface tension at the water-air interface on the inside of alveoli. Phospholipids are amphipathic molecules with a hydrophilic polar portion and a glycerol backbone. Fatty acids are attached to the glycerol backbone and this part of the molecule is hydrophobic. At a water-air interface, surfactant forms a monomolecular layer with the hydrophilic molecular heads associated with the water and the hydrophobic molecular tails exposed to the air (Figure 7.17). The monomolecular layer of surfactant causes the exclusion of water from the surface layer of the alveoli thus reducing the surface tension. Surfactant reduces the work of breathing and stabilizes alveoli during the ventilation cycle.

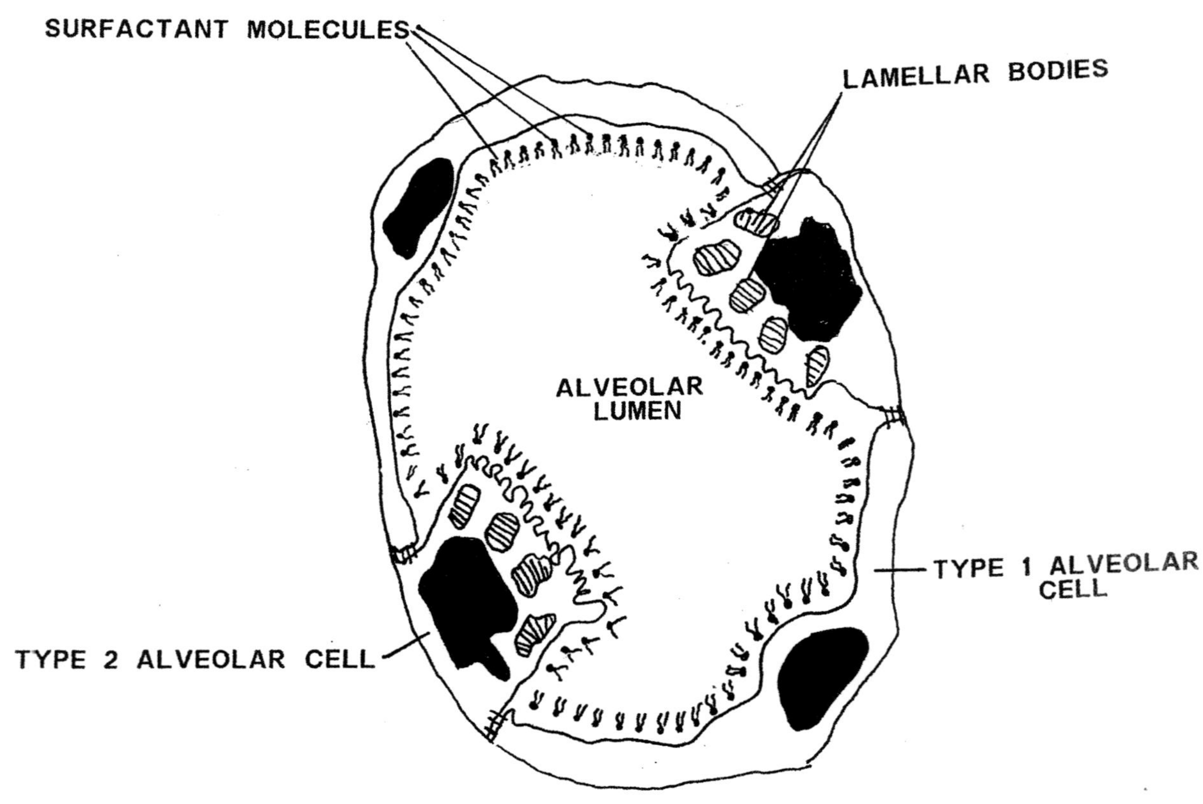

Figure 7.17. Surfactant monolayer on the alveolar epithelium.

MAMMALIAN FETAL DEVELOPMENT

Development of the Respiratory Membrane

The respiratory membrane is the tissue that separates the lumen of the alveolus from the lumen of pulmonary capillaries. In the fully developed lung, the respiratory membrane consists of an alveolar epithelium and its underlying basement membrane adjacent to the pulmonary capillary endothelium and its underlying basement membrane. The human respiratory membrane at midgestation is approximately 60 micrometers thick due to the presence of a large number of mesenchymal cells between the pulmonary capillaries and the alveolus. As gestation proceeds, the thinning of the interstitial tissues between the alveoli and pulmonary capillaries allows the basement membranes of the pulmonary capillaries and the alveolus to fuse. At birth, the average thickness of respiratory membranes in the human lung is approximately 0.2 micrometer.

Pulmonary Circulation

The development of pulmonary blood vessels occurs in parallel with the airways of the fetal lung. Most major pulmonary arteries are present in the human fetus by midgestation. Large pulmonary arteries contain considerable amounts of elastic tissue. Medium size pulmonary arteries have a large amount of smooth muscle tissue in the medial portion of the wall. Small pulmonary arteries and arterioles have little or no smooth muscle tissue in their walls, respectively.

The pulmonary vascular bed undergoes considerable growth during the last 3rd of gestation in humans and many large domestic animals. Most of the growth in the vascular bed is the result of growth and extension of the pulmonary capillaries. The growth and extension of the pulmonary capillaries dramatically increases the total cross-sectional area resulting in a gradual decrease in pulmonary vascular resistance and an increase in pulmonary blood flow as gestation proceeds.

DEVELOPMENT OF THE FETAL DIGESTIVE SYSTEM

Development of the Fetal Gastrointestinal Tract

By approximately the 10th week of gestation, the patent lumen of the human esophagus is lined by stratified ciliated columnar epithelium. Nonkeritinized stratified squamous epithelium can be found in the mucosa of the esophagus by the 5th month of gestation. The upper third of the human fetal esophagus is surrounded by skeletal muscle tissue that is well innervated by the nervous system by the 13th week of gestation. Swallowing in the human fetus can be observed as early as the 10th week of gestation. The lower third of the esophagus is surrounded by smooth muscle tissue. The middle third of the esophagus is surrounded by both skeletal and smooth muscle tissues. Esophageal and cardiomucous glands are observed in the human esophagus as early as the 28th week of gestation.

The fetal stomach contains tubular exocrine glands by the 18th week of gestation. As gestation continues, the glands increase in number faster than the gastric pits. Parietal cells are present in the fetal stomach by the 10th week of gestation but are not mature until approximately the 5th month of gestation. During fetal development, parietal cells produce little hydrochloric acid. Chief (zymogenic) cells containing zymogen granules are present in the stomach by the 11th week of gestation. The release of pepsinogen from the chief cells does occur during human fetal development. Goblet cells appear in the fundus of the human fetal stomach as early as the 11th week of gestation. Enteroendocrine (G) cells are present in the human fetal stomach as early as the 8th week of gestation.

MAMMALIAN FETAL DEVELOPMENT

The small intestine of the human fetus begins as an endodermal tube consisting of a single layer of epithelial cells. The epithelium of the endodermal tube rapidly stratifies and specific cell types of the small intestine such as enterocytes, goblet cells, and endocrine cells begin to appear. Ridges form from the underlying mesenchyme and then separate to form the villi of the small intestine. Terminally differentiated epithelial cells begin to cover the outer surface of the villi. Undifferentiated cells remain in clusters at the bottom of the villi. The lumen of the small intestine becomes patent and filled with fluids either swallowed or secreted from the gut. Changes in the length and diameter of the fetal small intestine stimulates undifferentiated cells in stratified areas to produce new villi. Villus cells produced in excess are shed at the tip of the villus during crypt growth.   Under the influence of cortisol, intestinal cell division and migration increases during late gestation. In some mammalian species this occurs between birth and weaning. By the 7th week of gestation, neurons appear in the midsection of the small intestine and neural development proceeds in a caudal direction reaching the rectum at approximately the 12th week of gestation. Auerbach's plexus [Leopold Auerbach, German anatomist, 1828 – 1897] develops between the 9th and 13th week of gestation. Auerbach's plexus is located between the inner and outer coats of the external muscularis and is responsible for peristaltic contraction in the intestine. Meissner's plexus [Georg Meissner, German histologist, 1829 - 1905] also develops between the 9th and 13th week of gestation and is located in the submucosa of the intestine. Meissner's plexus innervates the mucosa muscularis, lamina propria and epithelium. Besides contributing to peristaltic contraction in the small intestine, Meissner's plexus is also involved in stimulating segmental movements in the small intestine.

The large intestine of the human fetus also develops villi. Villi are predominant in the proximal area of the large intestine reaching maximal development at approximately the 5th month of gestation. As gestation continues, villi disappear and are replaced by crypts containing numerous goblet cells. Epithelial cells of the fetal colon are vacuolated suggesting these cells have the capacity to transport macromolecules.

Development of the Exocrine Pancreas of the Fetus

During the 12th week of human gestation, acini of the exocrine pancreas begin to develop from the blind ends of tubules that have branched and proliferated from the pancreatic diverticulum. The acinar epithelial cells become pyramidal in shape and contain zymogen granules by the 20th week of gestation.
During late fetal and postnatal development, the activity of the exocrine pancreas can vary considerable in mammalian species.

Fetal Gastrointestinal Motility

Human fetal swallowing begins in early development. The fetal human ingests from 300 ml to 1000 ml of fluid per day. Ingested fetal fluid consists of buccal secretions, liquids from the lung, and amnionic fluid. The intestinal tract and accessory organs also add fluid to the gastrointestinal tract. Water, electrolytes, glucose and other small molecules from fetal fluid are absorbed in the small intestine.   Swallowing of fetal fluid is important in the development of the tissues of the gut wall. This allows maternal substrates acquired by the fetus across the placenta to be used to develop other organ systems.

Peristaltic contractions can be observed in the human fetal stomach as early as the 14th week of gestation. Gastric emptying of the fetal stomach does not occur until approximately the 30th week of gestation. Weak intestinal peristalsis is observed in the human fetus at approximately the 11th week of gestation with migrating peristaltic contractions being observable by the 35th week of gestation.

Debris from fetal skin, large molecules in the amnionic fluid, sloughed intestinal cells, and bile pigments collect in the large intestine and form the **meconium**. The role of the meconium in gut development is not clear. The human meconium is usually eliminated shortly after birth. Since most fetal wastes are rid through the placenta, *in utero* defecation usually does not occur. The presence of low levels of gut enzymes in the amnionic fluid in the absence of a meconium indicates that a low level of intestinal secretion exist *in utero*.

<u>Absorption of Macromolecules and Cellular Digestion in the Fetal Small Intestine.</u>

Fetal enterocytes are found in all mammals either during fetal development or during the early neonatal period. Enterocytes have the ability to absorb and transport substances from the intestinal lumen (Figure 7. 18) and account for much of the digestion in the fetal small intestine. Pits located between adjacent villi on the apical surface of the enterocytes absorb material from the lumen by endocytosis into the apical endocytotic complex. After passing through the network of vesicular and tubular endosomes in the apical part of the cell, absorbed luminal material either goes to the supranuclear vacuole or to the lateral wall of the enterocyte. Absorbed material transported to the lateral wall of the enterocyte is released into the intercellular space. Nutrients that enter the supranuclear vacuole are degraded providing monomers for local development of gut tissues.

Although enterocytes account for most of the digestion in the fetal small intestine, some luminal digestion does occur. Disaccharidases and alkaline phosphatases are present on brushborder cells as early as the 10th week of gestation in the human fetus. Enterokinase appears at low levels in the human fetuses late in gestation. Luminal proteolysis is suppressed or absent during gestation.

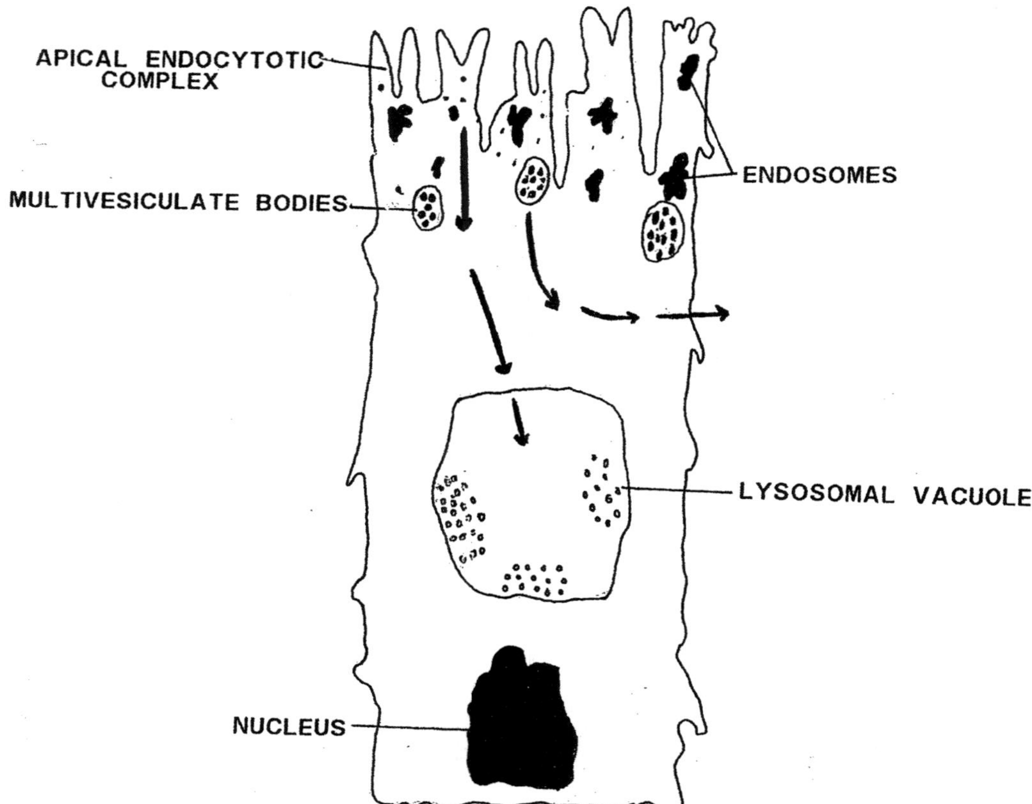

Figure 7.18. Human fetal enterocytes and the absorption of macromolecules from the intestinal lumen.

MAMMALIAN FETAL DEVELOPMENT

DEVELOPMENT OF THE FETAL URINARY SYSTEM

<u>Physiology of the Fetal Kidney</u>

Degeneration of the cloacal membrane from the 7th week of gestation to the 8th week of gestation allows the human fetus to excrete urine into the amnionic sac. During human gestation, the allantoic sac does not develop and the urine is excreted via the urethra only. In cattle, sheep and horses the fetal bladder has a second outlet that allows the urine to pass from the bladder, through a small tube in the umbilical cord, into the allantois.

At first, the secretion of urine in the human fetus is continuous, but becomes more periodic as the urinary bladder and its sphincter muscles develop. During late gestation, the bladder fills approximately every 30 minutes and is voided by contraction of the bladder wall during micturition. Human fetal urine production during late gestation averages 900 ml/day (0.2 ml/minute/kg of fetal body weight). The production of urine by the human fetus varies with fetal behavior and there is a 24-hour rhythm flow that may be related to maternal eating and drinking.

During late gestation the human fetus produces dilute urine (the osmolality of fetal urine is lower than that of fetal blood). The osmolality of human fetal urine ranges from 100 mOsm/kg of H_2O to 150 mOsm/Kg of H_2O compared to the osmolality of fetal blood plasma that is approximately 300 mOsm/Kg of H_2O. The production of dilute fetal urine occurs in response to an immature kidney. At the time of parturition the fetal kidney has not developed a high concentration of interstitial osmolite, and the medullary nephrons in the fetal kidney have not developed long loops of Henle that penetrate deep into the renal medulla.

In the human fetal kidney, 50% to 60 % of the filter ionic sodium is reabsorbed in the proximal convoluted tubule (compared to 80% to 85% in the adult kidney). The delivery of the excess ionic sodium to the distal convoluted tubule overwhelms the reabsorption mechanism. This allows approximately 7% of the filter ionic sodium to be excreted in the urine. Exogenous aldosterone has only modest influence of the excretion of ionic sodium in the fetal kidney. The high fetal excretion rate of ionic sodium and the high urine flow rate deliver a large volume of fluid to the amnionic sac thus promoting an adequate amnionic fluid volume.

DEVELOPMENT OF THE FETAL MUSCULAR SYSTEM

<u>Development of Muscle Fibers</u>

Muscle development in mammals is regulated by several myogenic factors. Some myogenic factors are required for the initial commitment and survival of **myoblasts** that are formed during embryonic development. Other myogenic factors are required for the differentiation and fusion of myoblasts into multinucleated myotubes.

Myotubes transcribe the muscle-specific contractile proteins and assemble them into the arrays of thick and thin myofilaments that slide in relation to each other. Thin filaments are formed by the coordinated accumulation of troponin, tropomyosin and actin. Thick filaments are formed by the polymerization of myosin. The heavy and light chains of myosin exist as variants so that a large number of recombinants can be produced. Myosin isoforms differ in the rate in which they split ATP, thus determining the rate by which a muscle fiber contracts.

Efferent Neural Innervation of Skeletal Muscle Fibers

The growth cone of nerve fibers that grow out of the dorsal root ganglia and the ventral root of the spinal cord are directed to clumps of mesodermal cells (by substances found in the extracellular matrix) that will differentiate into the definitive muscles. The stage of contact between efferent nerve fibers and primitive muscles usually occurs at the time when myoblasts have fused to form multinucleated tubes. The acetylcholine receptors on postsynaptic myotubes are initially spread over the entire surface (Figure 7.19). Efferent nerve fibers can innervate a primitive muscle fiber at many points along its length. Numerous nerve fibers can also innervate a primitive muscle fiber at a given point along its length. As gestation proceeds, this hyperinnervation of primitive muscle fibers is gradually reduced by the removal of all but a single axon from the site of synapse formation. Acetylcholine receptors on the postsynaptic myotube remote from neural contact disappear or migrate to areas of the cell membrane where the motor end plate is formed. The terminal arborization of the surviving contact also becomes more complex. The neuromusclular junction may be strengthened and stabilized by various tropic substances. Muscular tropic substances are absorbed by the nerve terminal and then transported to the soma of the efferent nerve fiber.

Afferent Innervation of Skeletal Muscle Fibers

Afferent nerve fibers from the dorsal root ganglia reach the myotube at approximately the same time as the efferent nerve fibers. It is not clear whether afferent nerve fibers seek out myotubes destined to become intrafusal fibers or whether myotubes in contact with afferent nerve fibers are stimulated to become intrafusal muscle fibers. The point of contact of the myotube and the sensory nerve fiber becomes encapsulated by connective tissue derived from the perineurium. Within the encapsulation, various types of muscle spindles form and generate action potentials relaying information to the central nervous system on changes in muscle length. Afferent nerve fibers also innervate the tendon region of primitive muscle fibers resulting in the encapsulation of this region. Within this encapsulation, the Golgi tendon organ [Camillo Golgi, Italian histologist, 1843 – 1926] is formed relaying information to the central nervous system concerning load on the muscle.

Changes of the Development of Skeletal Muscles Fibers during Gestation

The development and maintenance of skeletal muscle fibers is dependent on efferent nerve innervation. Denervation of the efferent nerve fiber results in cessation of growth and maturation, and atrophy of the muscle. The efferent nerve fiber is believed to release tropic factors that are absorbed by the muscle fiber. These tropic factors are responsible for maintaining the normal growth and metabolic processes within a muscle fiber. These neurotropic factors are probably small diffusable polypeptides acting on specific receptors on the myoblast, myotube or primitive muscle fiber.

As human gestation proceeds, embryonic myosin is slowly replaced by fetal (the presence of fetal or neonatal myosin, depends on the mammalian species and the length of gestation) myosin. Fetal and neonatal myosins are slow myosins and are responsible for the slow contractions observed in fetal and neonatal muscle fibers. Factors regulating the expression of myosin genes in muscle fibers are not completely understood. Comparison of efferent nerves innervating fast twitch muscle fibers to those nerve fibers innervating slow twitch muscle fibers reveals that there is a difference in the pattern of bioelectrical

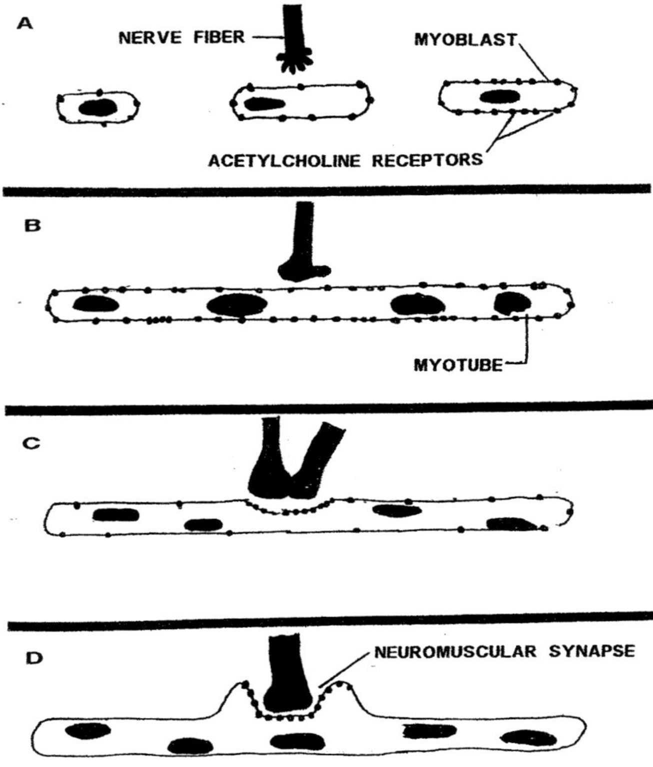

Figure 7.19. Formation of the neuromuscular synapse. (A) Arrival of nerve fibers at the surface of the myoblast. (B) Formation of a multinucleated myotube and dispersed expression of acetylcholine receptors. © Multiple neural innervations and acccumulation of postsynaptic acetylcholine receptors at the synapse. (D) Multiple innervations are eliminated.

activity is transmitted to the muscle fibers. Slow twitch muscle fibers receive an almost continuous pattern of discharge from its innervating efferent nerve fiber. Fast twitch muscle fibers receive bursts of neural discharge at relatively high frequencies from its innervating efferent nerve fiber. This difference in the pattern of bioelectrical activity being transmitted by efferent nerve fibers is probably due to the different excitatory and inhibitory inputs being received by the efferent nerve soma from higher levels of the central nervous system. In "cross-innervation" experiments, when a slow muscle fiber is innervated by a foreign fast nerve fiber, the muscle fiber expresses fast myosin and physiologically becomes a fast twitch fiber. Fast twitch muscle fibers that are experimentally innervated by slow nerve fibers increase the expression of slow myosin and thus contract more slowly. This suggests that the expression of a particular myosin isoform may be partially dependent on the interaction between the muscle fiber and a particular type of efferent nerve fiber.

Most muscles of the mammalian body must have the ability to maintain prolonged contractions as well as have the ability to contract briefly and often. These muscles must have slow twitch muscle fibers and fast twitch muscle fibers interdigitated within the muscle or distributed across the muscle to increase the efficiency of each function. How muscles in mammals are brought to their complete functional development is poorly understood.

MAMMALIAN FETAL DEVELOPMENT

DEVELOPMENT OF THE FETAL REPRODUCTIVE SYSTEM

Introduction

The development of the reproductive organs in eutherian mammals begins with the establishment of genetic sex. **Genetic sex** is determined in eutherian mammals when either an X chromosome carrying spermatozoa or a Y chromosome carrying spermatozoa fertilizes the secondary oocyte. **Gonadal sex** is the determination of the fate of the bipotential gonads. In XY embryos, bipotential gonads develop into testes, whereas the bipotential gonads in XX embryos develop into the ovaries.

Gonadal Sex Differentiation and Sex Determination

Human gonads develop from the urogenital ridge at approximately the 4th week of gestation. Germ cells migrate from the yolk sac to the gonadal ridge. The remainder of gonadal cells is derived from the mesoderm of the urogenital ridge. The formation of the bipotential gonad in humans is complete by the 5th week of gestation and consists of support cells (the eventually give rise to Sertoli cells in the male and granulosa cells in the female), stromal cells and germ cells. The appearance of primordial Sertoli cells and the development of seminiferous tubules is the first histological sign of sexual dimorphism of the gonads. Leydig cells develop from mesenchymal cells at approximately the 8th week of gestation. The primitive ovary consists primarily of mesonephric cells with clusters of oogonia found at the periphery. Follicular cells surround oogonia after meiotic arrest.

Two genes called steroidogenic factor 1 (SF1) and Wilm's Tumour 1 (WT1) [Max Wilms, German surgeon, 1867 – 1918] primarily regulate the development of the bipotential gonad in humans. After formation of the bipotential gonad, the sex-determining region of the Y chromosome (SRY) regulates testicular development in XY individuals. The sex-determining region of the Y chromosome (SRY) suppresses the activity of the DAX 1 gene. This removes the repressive influence of DAX 1 on SF1. Expression of SF1 results in the activation of other target genes including **Mullerian inhibiting hormone (MIH)** [Johannes P. Muller, German anatomist, physiologist and pathologist, 1801 – 1858], SOX 9 ((sex determining region Y)-box 9) and enzymes involved in the synthesis of androgens. Mullerian inhibiting hormone is responsible for the regression of the Mullerian ducts that eventually would give rise to the female reproductive tract. Testosterone masculinizes the Wolffian ducts to form the male reproductive tract. In XX embryos, the lack of SRY means that DAX 1 continues to suppress SF1 preventing the activation of MIH and the enzymes involved in the synthesis of androgens. This allows the Mullerian ducts to develop into the female reproductive system.

Differentiation of the Central Nervous System

Sexual differentiation of the central nervous system is caused by the release of steroids from the developing gonads. Sexual differentiation of the central nervous system in the rat and the guinea pig occurs from day 18 to day 27 and from day 30 to day 37, respectively. Testosterone produced by the fetal testes is transported to the brain. The brain contains aromatase that converts testosterone to estradiol 17-*B*. Estradiol 17-*B* is responsible for sexual differentiation of specific areas of the brain by altering the rate of cellular apoptosis and inducing the production of neurotropins that promote cell survival.

Sexually differentiated regions of the brain are mostly contained in the hypothalamus. Sexual behavior is controlled by the sexually dimorphic nucleus of the preoptic area. The sexually dimorphic nucleus of the preoptic area of the hypothalamus is larger in males when compared to that observed in the female. Gender differences in the secretion of gonadotropins are controlled by the anteroventricular periventricular nucleus. The anteroventricular periventricular nucleus of the hypothalamus is smaller in males when compared to that of the female.

<u>Development of the Hypothalamus and the Pituitary</u>

The development of the human hypothalamus and pituitary are coordinated yielding a system that regulates endocrine function. Many neuron cell bodies are found within specific nuclei in the hypothalamus. Neurons producing CRF are found in the paraventricular nucleus. The neurons producing GnRH are more diffusely located in a region called the preoptic area. Neurons that secrete GnRH arise in the olfactory placode and migrate along the nervus terminalis-vomeronasal complex to the preoptic area of the hypothalamus. Rapid mitosis of these future GnRH secreting cells occurs and their fibers grow into the median eminence. The adenohypophysis develops from a structure called Rathke's pouch. Five specialized types of cells of the adenohypophysis develop from a common progenitor cell in a spatial and temporal sequence. Corticotrophs are the first cells of the adenohypophysis to develop, followed by the thyrotrophs and gonadotrophs. The lactotropes and the somatotropes of the adenohypophysis are the last to develop.

The secretion of plasma gonadotropins during human fetal development peaks at approximately midgestation, declines in the last trimester rises again after birth for a few months before declining to low levels until the onset of puberty. The plasma concentration of FSH and LH at midgestation is lower in males compared to that observed in the female. In mammalian gonadal development, specific time periods occur by which gonadotropins can influence gonadal development. In the male lamb and rat, Sertoli cell proliferation ceases after the 10th week of postnatal life and after the 15th day of postnatal life, respectively. Sertoli cell proliferation in the human testes is believed to occur throughout most of prepubertal development. Treatment of males with gonadotropins during periods of Sertoli cell proliferation can increase the number of cells within the testes. Treatment of males with gonadotropins outside the period of Sertoli cell proliferation has no influence on Sertoli cell number. Disruptive changes of the pattern of gonadotropin secretion during the critical window of Sertoli cell proliferation may impair testicular development and the future reproductive performance of the male.

Internal mechanisms regulating the secretion of fetal gonadotropic hormones is poorly understood. Testosterone secreted from the testes is responsible for causing the sexual differentiation in the secretion of gonadotropins. The concentration and pulse frequency of LH of the human male fetus in late gestation is reduced compared to values observed in the female. It is possible that steroids (e.g. estrogen) from the placenta rather than the gonads take over the suppression of gonadotropin secretion.

Environmental factors can also influence gonadal development of the mammalian fetus. Changes in daylight can be relayed to the fetus by the transplacental transfer of maternal melatonin. Environmental estrogen-like substances can mimic the effects of estrogen by binding to estrogen receptors and disrupting normal fetal reproductive development. Exposure of pregnant females to octylphenol results in a decrease in the secretion of FSH in the developing fetus. If pregnant females are exposed to octylphenol during a period of Sertoli cell proliferation in the developing male fetus, the decrease in plasma FSH concentration would suppress Sertoli cell proliferation resulting in reduced testes size and lower sperm counts at maturity.

MAMMALIAN FETAL DEVELOPMENT

FETAL METABOLISM

Glucose and Carbohydrates

Glucose is the primary source of energy for the developing mammalian fetus. The developing fetus has little capacity for gluconeogenesis because gluconeogenic enzymes are essentially inactive in the presence of low oxygen levels. The fetus obtains most of its glucose from the maternal blood. Progesterone increases maternal appetite and stimulates maternal glucose deposition as fat during early gestation. As gestation progresses, hCS mobilizes fatty acids from maternal fat stores. Fatty acids are important in maternal metabolism because maternal tissues become less sensitive to insulin as gestation progresses. Fatty acids of maternal origin can also be transferred across the placenta to the fetus.

The rate of fetal glucose utilization is dependent on insulin from the fetal pancreas. If the fetal plasma concentration of glucose is elevated (as in gestational diabetes), fetal growth and fat storage are promoted and overweight neonates are usually the result. High concentrations of cardiac glycogen allow the heart to maintain its contractility in the presence of hypoxia. The fetal brain has no glycogen stores and is totally dependent on blood glucose for energy. This explains why hypoglycemia accompanied with hypoxia is detrimental to fetal brain development and function. The progressive influence of the fetal adrenal cortex near term is important in the deposition of glycogen in the liver. The storage of glucose as fat within the fetus is regulated by insulin. When plasma glucose concentration and growth requirements of the fetus are being maintained, fetal insulin converts glucose into fat. Brown fat, a thermogenic tissue, found in the human fetus and neonate can generate heat independent of muscle contraction and shivering.

Fatty Acids

There is a great demand for long-chain polyunsaturated fatty acids by the human fetus during the last third of gestation. Arachidonic acid and docosahexanoic acid are important components of the structural lipids of the developing fetal brain. The placenta is essentially impermeable to lipids. The fetal demand for lipids is therefore met by both the placental transfer and endogenesis production of fatty acids. The order of selectivity of in the placental transfer of fatty acids to fetal circulation is docosahexanoic acid, alpha-linoleic, linoleic, oleic and arachidonic.

Amino Acids and Urea

Maternal intermediate metabolism of amino acids increases during gestation when compared to the nongravid state. The maternal excretion of urea decreases during pregnancy indicating that dietary amino acids are being used more efficiently. Elevated levels of progesterone during pregnancy decrease the ability of the maternal liver to deaminate amino acids. Accumulated maternal amino acids are transported across the placenta to the fetus. Since maternal blood has lower levels of endogenous urea, fetal urea readily diffuses across the placenta into the maternal blood.

Calcium and Phosphorus

There is a great demand for calcium and phosphorus during the last third of mammalian gestation when considerable ossification is occurring. Approximately 23 grams of calcium and 13.5 grams of phosphorus accumulate in the human fetus during gestation. The total amount of calcium and phosphorus needed by the human fetus represents only 2% of that found in maternal bone tissue. Normally the average daily intake of calcium greatly exceeds adequacy. Calcium and phosphorus are efficiently transported from maternal circulation to fetal circulation by an active transport mechanism. Therefore, the fetus is a minimal drain on maternal calcium and phosphorus stores thus allowing a positive calcium and phosphorus balance to be maintained during pregnancy.

<u>Iron</u>

Free and bound iron can be found in both maternal and fetal blood during mammalian gestation. Plasma iron levels in the human fetus are approximately 2 to 3 times greater than that found in maternal blood due to a greater concentration of unbound iron accumulated by active placental transport. The human fetus usually requires approximately 300 mg of iron and the placenta requires approximately 50 mg of iron. Approximately two-thirds of the fetal iron is incorporated in the form of hemoglobin, which is being produced as early as the third week of gestation. The remaining one third of the iron is stored in the liver. This iron is used to synthesize additional hemoglobin that will be used after birth. Approximately 500 mg of iron is required to increase the maternal mass of hemoglobin and approximately 200 mg of iron is required to counterbalance the loss of iron at parturition. Maternal iron absorption increases over nongravid levels from approximately 10% in the first trimester to approximately 30% in the last trimester of human pregnancy. However, during human pregnancy there is a 40% to 90% incidence of maternal serum iron deficiency. Therefore iron supplements are usually taken during human gestation.

<u>Vitamins</u>

Vitamins function in the fetus in the same manner as that observed in the adult. Vitamin B_{12} and folic acid are provided to the developing mammalian fetus at the expense of maternal stores. Folic acid is important in nucleoprotein synthesis, amino acid metabolism and fatty acid metabolism. Vitamin B_{12} is needed in the formation of erythrocytes. Vitamin B_{12} is slowly absorbed across the ileum of the maternal small intestine and is transported in the maternal blood by transcobalamin II. Although maternal vitamin B_{12} falls to minimum levels at approximately the 16th to the 20th week of human gestation, this is still high enough to meet fetal demands. Vitamin C is needed for the formation of protein fibers in connective tissue as well as the formation of intercellular substances in bone matrix. Vitamin D is needed for normal fetal bone development, but is more important in the mother for the absorption of calcium from the gastrointestinal tract. If the mother has more than adequate levels of vitamin D, large quantities of this vitamin will be stored in the fetal liver to be used by the newborn for several months after birth. The exact role of vitamin E in fetal development is unclear. Removal of vitamin E from the diets of rats results in spontaneous abortion of fetuses during early gestation. Vitamin K is produced by the microflora of the maternal gastrointestinal tract. Vitamin K crosses the placenta and is used by the fetal liver for the formation of factor VII. At birth, the newborn has no adequate source of vitamin K until the intestinal microflora becomes established after the first week of life. Therefore, prenatal storage of small quantities of vitamin K in the fetal liver is helpful in preventing hemorrhaging for a short time after birth.

<u>Bilirubin</u>

Bilirubin can be described as being made from the heme moiety of hemoglobin without the iron. Uridine diphosphate glucuronyl transferase is not functional in the human fetal liver until late pregnancy or early neonatal development. Therefore the fetal liver cannot convert bilirubin to bilirubin glucuronide for its incorporation into bile. Fetal hemoglobin is degraded in the bone marrow to form bilirubin. To rid the fetus of bilirubin, bilirubin is attached to serum albumins and transported to the placenta (Figure 7.20). Unconjugated fetal bilirubin then diffuses (down its concentration gradient) across the placenta to the maternal blood. Unconjugated bilirubin from the fetus then attaches to serum albumins in the maternal blood and is transported to the liver. Fetal bilirubin is then converted to bilirubin glucuronide in the maternal liver and secreted into the maternal small intestine as bile.

MAMMALIAN FETAL DEVELOPMENT

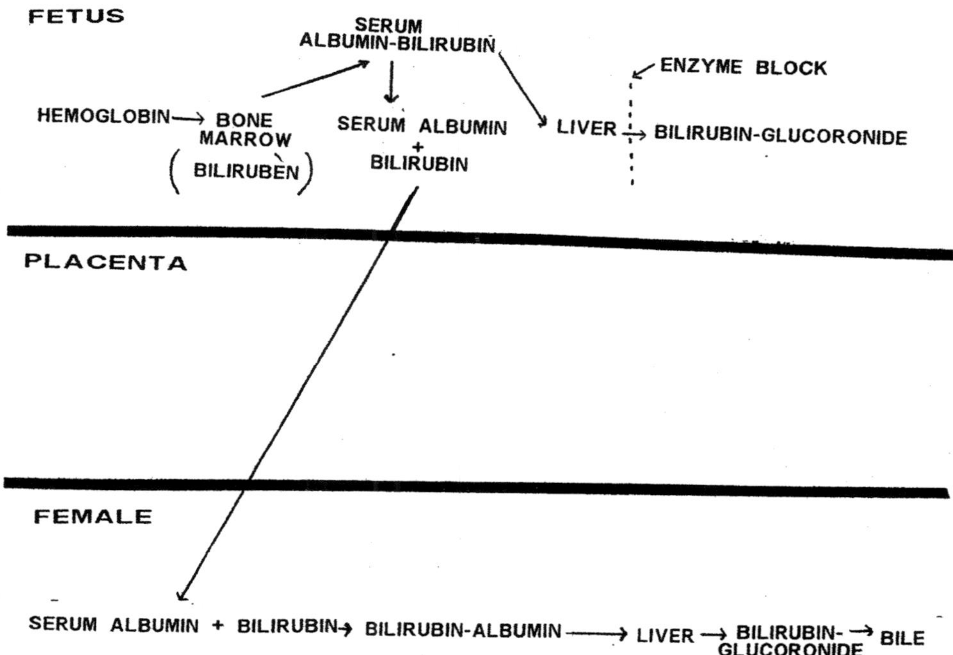

Figure 7.20. The metabolism of fetal bilirubin during human gestation.

FETAL SURVIVAL AND MATERNAL IMMUNE RESPONSES

<u>Antigenicity of the Conceptus and Protective Immunological Barriers</u>

Histocompatability antigens are present in small amounts on the human embryo shortly after implantation. Although the syncytiotrophoblast and cytotrophoblast appear to lack class I and class II histocompatability antigens, trophoblastic cells invading the uterus do express histocompatability antigen C and histocompatability antigen G that is not expressed by other fetal and maternal cells.

Placental anatomy of many mammalian species prevents the passage of maternal cells to the fetus. In monkeys and humans, maternal IgG normally crosses the placenta from maternal circulation to fetal circulation by transport mechanisms. This gives passive immunity to the fetus and neonate.
However, passage of antibodies across the hemochorial placenta can result in detrimental conditions for the developing fetus and neonate. **Erythroblastosis fetalis** (hemolytic disease of newborns) is a condition that is the result of an Rh incompatibility between the mother and the developing fetus. If the fetus inherits the Rh dominant gene from its father and the mother is Rh negative (homozygous recessive), fetal hemorrhaging during the first gestation and parturition can sensitize the maternal immune system to produce antibodies and memory cells to the Rh factor. Exposure of the mother to the Rh factor during the second and subsequent pregnancies results in antibodies to the Rh factor crossing the placenta and

rupturing of fetal erythrocytes. In most cases, the presence of an antibody in fetal circulation to a foreign antigen would be harmlessly disposed of by free soluble antigens. Any remaining antibodies would be distributed among a wide range of cell types and thus be diluted. Body cells would only bind just a few antibodies and would not be lysed. In erythroblastosis fetalis, each fetal erythrocyte can bind numerous antibodies to the Rh factor. This results in lysis of the fetal erythrocytes. Before the advent of modern obstetrics, this condition occurred in mild to serious forms in every 1:50 to 1:100 newborns. Today this conditioned is remedied by the administration of **Rho-Gam** during gestation and parturition. Rho-Gam binds the antigen of the fetal red blood cells so the mother's immune system does not become sensitized.

8

MAMMALIAN PARTURITION

INTRODUCTION

Labor is defined as regular uterine contractions resulting in the thinning and dilation of the cervix so the product of conception can pass out of the uterus. Human labor can be divided into 4 phase's referred to as pregnancy (phase 0), uterine activation (phase 1), uterine stimulation (phase 2) and involution of the uterus (phase 3). Each of these phases and the mechanisms involved will be discussed below.

PREGNANCY

<u>Endocrinology of Late Mammalian Gestation</u>

Although there are a few low frequency contractions of the mammalian uterine musculature during **pregnancy (phase 0)**, the uterine musculature is essentially quiescent. The myometrium is essentially unresponsive to uterotonic stimulants (e.g. prostaglandins and oxytocin). Quiescence of the uterine musculature during pregnancy appears to involve the blockage of gene expression involved in increasing myometrial activity and the synthesis of myometrial inhibitors.

Inhibitors of uterine muscular contraction include progesterone, prostacyclin, nitric oxide, parathyroid related peptide, calcitonin gene regulated peptide, relaxin, adrenomedullin and vasoactive intestinal peptide. Increased levels of progesterone during pregnancy inhibit both myometrial contractility and the formation of gap junctions between smooth muscles cells of the myometrium. Progesterone prevents the expression of genes involved in the formation of receptors for both oxytocin and prostaglandin thereby inhibiting myometrial response to endogenous oxytocin and prostaglandins. **Parathyroid hormone related peptide (PTHrP)** is an autocrine/paracrine factor produced by the myometrium. Parathyroid hormone related peptide increases intracellular c-AMP by activation of the alpha subunit of the G protein complex. Myometrial stretching increases gene expression of the PTHrP in rodents. This appears to maintain quiescence of the myometrium when the fetus increases in size during the last third of gestation. **Nitric oxide** is an endogenous muscle relaxant and appears to work synergistically with progesterone to maintain myometrial quiescence. In rats, a decrease in the production of nitric oxide by decidual and myometrial tissues occurs prior to parturition presumably removing the inhibitory influence on myometrial quiescence. A corresponding increase in levels of nitric oxide by inflammatory cells of the cervix indicates a possible role in cervical effacement and relaxation. Relaxin has a dual role in mammalian reproduction in inhibiting myometrial contraction during gestation and regulating changes in the connective tissue in the cervix. Relaxin suppresses spontaneous myometrial contractions and inhibits the effects of oxytocin via a secondary messenger system in both the guinea pig and rat. The up-regulation of **matrix metalloproteinase 1 (MMP-1)** and **matrix metalloproteinase 3 (MMP-3)** gene expression in the cervix by relaxin inhibits **tissue inhibitor of metalloproteinase 1 (TIMP-1)** allowing remodeling of the connective tissue of the cervix.

MAMMALIAN PARTURITION

UTERINE ACTIVATION

Contraction-Associated Proteins in Uterine Activation

Uterine activation (phase 1) occurs as mammalian gestation nears completion. Uterine activation involves the expression of a number of genes involved in the synthesis of **contraction-associated proteins (CAP's)** that prepare the myometrium of the uterus for contraction.   The activation of the CAP's involves both endocrine and mechanical changes.

As pregnancy nears completion, the expression of the CAP genes increases (Figure 8.1). The number of myometrial receptors for both prostaglandin and oxytocin increases. The number of myometrial ionic sodium channels increases allowing for greater depolarization. An increase in the number of ionic calcium channels on myometrial cells allows for activation of the contractile machinery.   In increase in myometrial cell to cell coupling allows for greater synchronization of the waves of contraction.

Endocrine Changes Involved in Uterine Activation

Uterine activation in many mammalian species requires the removal of the effects of progesterone on the myometrium. In many nonprimate mammals this is accomplished by a decrease in the maternal plasma levels of progesterone thus increasing ratio of estrogen to progesterone. The fetal hypothalamus produces and secretes CRH that is transported to the fetal adenohypophysis by the hypothalamic-hypophyseal portal system. Stimulation of the fetal adenohypophysis by CRH increases the synthesis and secretion of fetal ACTH. Fetal ACTH stimulates the fetal adrenal glands to produce and release cortisol. Fetal cortisol is transported to the placenta.   Increased levels of fetal cortisol induce the activity of placental enzymes 17 alpha-hydroxylase, steroid C_{17} and C_{20} lyase and probably aromatase. The activation of these enzymes diverts placental progesterone conversion to estrogen synthesis thus increasing the maternal estrogen/progesterone ratio in the blood (Figure 8.2)

The ovine fetus has evolved several mechanisms for preventing the negative feedback of cortisol on the secretion of CRH and ACTH by the fetal hypothalamus and fetal pituitary, respectively. First, fetal cortisol stimulates the synthesis and secretion of **corticosteroid binding globulin**. Corticosteroid binding globulin binds circulating cortisol preventing its negative feedback effects. Secondly, the ovine fetal pituitary expresses increased amounts of 11-*B* hydroxysteroid dehydrogenase that has the ability to inactivate circulating cortisol by its oxidation to cortisone. Lastly, the periventricular nucleus of the fetal hypothalamus has lower number of cortisol receptors compared to that of the hypophysis.

Pregnancy in the goat is entirely dependent on the corpus luteum. At the time of parturition the adrenal cortex of the fetal goat produces glucocorticoids that stimulate the placenta to convert dehydroepiandrosterone and dihydro-3-epiandrosterone sulphate to estrogens. Elevated placental estrogens enhance the production of prostaglandin $F_{2\,alpha}$. Prostaglandin $F_{2\,alpha}$ causes regression of the corpus luteum resulting in a large decrease in maternal plasma levels of progesterone. The drop in maternal plasma progesterone concentration increases the estrogen: progesterone ratio.

The endocrinology of uterine activation in nonhuman primates as well as humans is poorly understood. Human maternal plasma levels of progesterone do not decrease at the time of parturition as compared to other mammalian species. An increase in placental synthesis of estrogen does occur near the time of parturition. During the last few weeks of pregnancy increased fetal adrenal activity increases the level of cortisol in fetal plasma. Cortisol cannot produce estrogen from placental progesterone because the human placenta lacks 17 alpha-hydrolase activity. Cortisol does stimulate the secretion of placental CRH.

MAMMALIAN PARTURITION

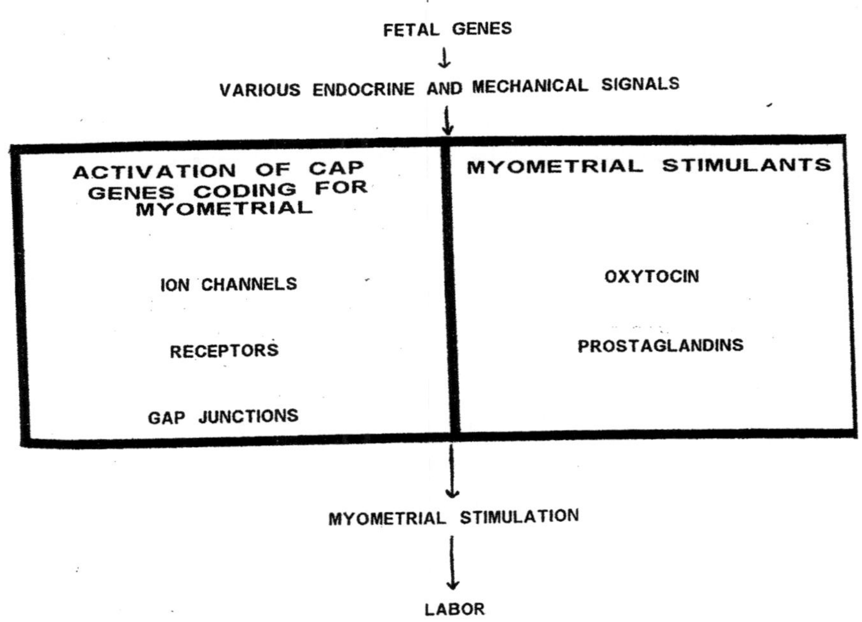

Figure 8.1. General outline of mammalian uterine activation at parturition.

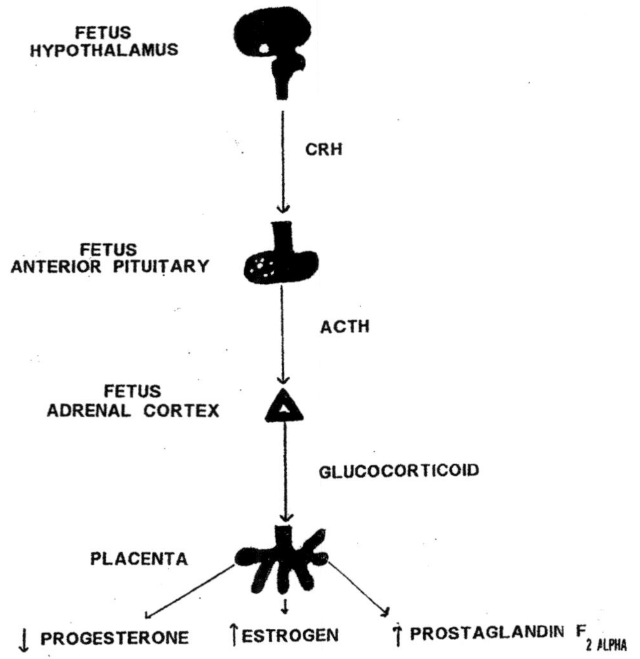

Figure 8.2. General model of the endocrinology of uterine activation in nonprimate mammals.

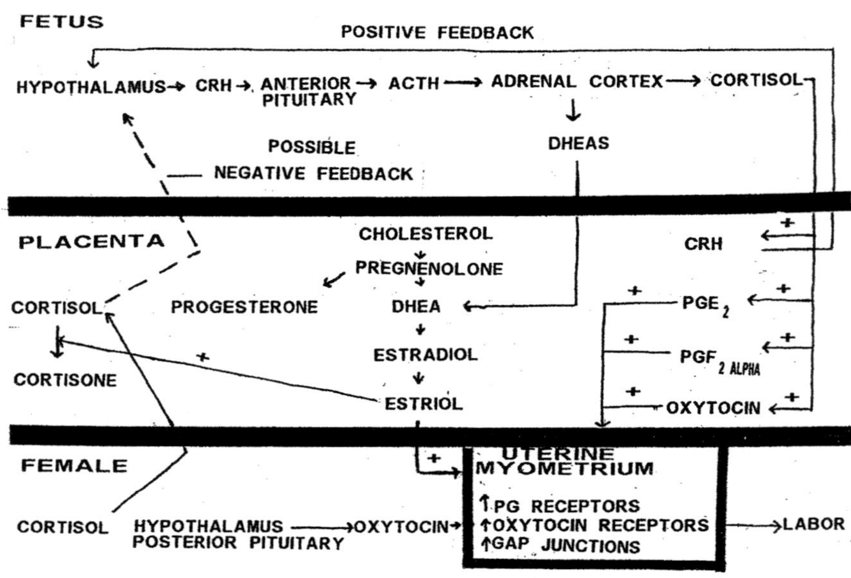

Figure 8.3. General model of endocrinology of uterine activation in humans and other primates. Dehydroepiandrosterone (DHEA), Dehydroepiandrosterone sulfate (DHEAS), and activation (+).

Placental CRH appears to be responsible for the increased fetal adrenal production of androstenedione. Placental CRH appears to be also responsible for increase in placental aromatase activity converting androstenedione to estrogen (Figure 8.3). Placental CRH has other multiple sites of action that include stimulating the fetal pituitary to produce ACTH. Placental CRH also effects the uterus by inducing placental vasodilation, prostaglandin production and increase myometrial contraction.

<u>Mechanical Changes Involved in Uterine Activation</u>

There are 3 phases of uterine growth referred to as hyperplasia, hypertrophy and plateau. **Uterine hyperplasia** occurs during the first third of mammalian gestation and occurs primarily in response to endocrine regulation. **Uterine hypertrophy** parallels fetal growth during the last third of gestation and is progesterone dependent. Hypertrophy of the uterus occurs in response to tension placed on the myometrium in response to fetal growth. The **plateau phase of uterine growth** is characterized by cessation of uterine hypertrophy even though the fetus continues to grow.

Mechanical stretching of the human uterine myometrium enhances uterine excitability during late gestation. Human twins are born on the average of nineteen days earlier than a single child. Medical doctors can induce labor in women by rupturing the fetal membranes allowing the fetal head to move into the cervix. The mechanism by which stretching of the human uterine myometrium increases uterine excitability is unknown. Studies in rats suggest that **focal adhesions** may serve as mechanotransducers by which uterine stretching can increase uterine excitability. Focal adhesions can be described as "spot welds" of myometrial cells to it's surrounding matrix. Focal adhesions are formed from an aggregation of integrin molecules that traverse the plasma membrane of myometrial cells.   The intracellular domains of the integrin molecules serve as a platform to which a number of adaptor proteins bind. These adaptor proteins allow the integrin molecules to be anchored to cytoskeletal actin. Some of the intracellular adaptor proteins contribute to the activation of intracellular signaling pathways that are involved in the regulation of gene expression. The extracellular domain of the integrin molecules is bound to components of the extracellular matrix. The turnover of focal adhesions is dependent on **focal adhesion kinase**. The expression and activity of focal adhesion kinase is low at midgestation, but increases markedly by late gestation. During late pregnancy, focal adhesion kinase permits the remodeling of focal adhesions. This allows the hypertrophy of myocytes within the myometrium to keep pace with fetal growth preventing the activation of CAP gene expression. A decrease in progesterone causes a decrease in the activity of focal adhesion kinase preventing the remodeling of focal adhesions. This results in the formation of large focal adhesions that anchor muscle cells to the extracellular matrix. Myocyte hypertrophy ceases and continued fetal growth increases the tension on the myocytes. Increased myometrial tension is detected by the focal adhesions that then activate the CAP genes and initiating the onset of labor.

UTERINE STIMULATION

Prostaglandins and Uterine Stimulation

Uterine stimulation (phase II) is the period of parturition when the primed uterus is contracting in response to uterotonic substances. Prostaglandins and oxytocin are uterotonic in many mammalian species.

Prostaglandin E_2 and prostaglandin $F_{2\text{ alpha}}$ are the primary stimulatory prostaglandins involved in mammalian parturition. Primate uterine tissue is high in arachidonic acid at the time of parturition. Arachidonic acid in the decidua and fetal membranes serves as a precursor to the synthesis of prostaglandins. **Cyclo-oxygenase enzyme 1 (COX-1)** and **cyclo-oxygenase enzyme 2 (COX-2)** are both expressed in the uterine tissue with COX-2 being sensitive to glucocorticoid induction. Prior to labor, prostaglandins increase in maternal plasma, urine and amnionic fluid. Prostaglandins induce cervical ripening, myometrial contractions and increased myometrial sensitivity to oxytocin. Prostaglandins increase the influx of ionic calcium across smooth muscle cell membranes, enhance the release of intracellular stores of ionic calcium and increase myometrial gap junction formation.

Oxytocin and Uterine Stimulation

Oxytocin found in maternal blood during late pregnancy is primarily derived from the fetus, fetal membranes and placenta. During the second half of pregnancy there is an increase in the numbers of myometrial oxytocin receptors that increases myometrial sensitivity to oxytocin. Oxytocin attaches to receptors on myometrial cells activating G protein that results in the increase in the levels of intracellular ionic calcium resulting in contraction.

Oxytocin appears to be important in determining the time of day that parturition occurs. Prior to the onset of recognizable labor, myometrial activity switches from contractures to contraction. Labor proceeds when the female is hidden and safe from predators. The switch from contractures to contractions in diurnally active animals occurs at night. The switch from contracture to contractions in nocturnally active animals occurs during daylight. This circadian rhythm of uterine activity is accomplished with an increase in the maternal levels of circulating oxytocin and myometrial oxytocin receptors. Oxytocin induces prostaglandin synthesis and the formation of gap junctions in the myometrium. This suggests that oxytocin may work in synergy with other factors to initiate parturition.

Mechanics of Human Parturition

Once labor has commenced, pain signals from the uterus and the vagina elicit neurogenic reflexes from the spinal cord to the abdominal muscles. Excitation and contraction of the abdominal muscles enhance the process of fetal expulsion.

The first stage in activation of the human uterus is progressive cervical dilation. This stage usually lasts between 8 hours to 24 hours in the female's first pregnancy, but decreases considerably in subsequent pregnancies. The second stage of uterine activation allows the fetal head to move through the birth canal until parturition is complete. This stage may last one half hour or more in the female's first pregnancy, but decreases to only a few minutes in subsequent pregnancies.

The human myometrium undergoes periodic episodes of weak rhythmic contractions during gestation. During parturition, contractions begin at the superior end of the uterus and spread downward over the uterus. The strength of the contractions is greater at the fundus and body and weakens at the caudal end adjacent to the cervix. Prior to labor, PGE causes contraction of the myometrium in the fundal and lower areas of the uterus. During labor, PGE causes myometrial contraction in the fundus but relaxation in the lower uterus. This may allow decent of the fetus into the birth canal. Contractions become exceptionally strong during fetal expulsion. The reason for the increase in the strength of uterine contractions during fetal expulsion is unknown. A theory has been proposed based on a positive feedback mechanism. This theory suggests that stretching of the cervix by the fetal head evoke a strong reflex that increases the contractility of the uterine body and pushes the fetus forward. Movement of the fetus further stretches the cervix thus enhancing the positive feedback to the uterine body. This feedback mechanism continues until the fetus is expelled from the uterus. Several facts appear to support the positive feedback theory for human parturition. First, once the strength of uterine contractions become greater than a critical threshold, each contraction leads to subsequent stronger contraction until the maximum effect is achieved. Secondly, cervical stretching due to the fetal head causes the entire corpus of the uterus to contract pushing the fetus further down the cervix. The positive feedback theory for parturition can also be used to possibly explain false labor. After labor starts, if any uterine contraction falls below the critical threshold value, the positive feedback cycle may go into a retrograde decline causing labor contractions to eventually fade away.

At the onset of labor, uterine contractions may be 30 minutes apart. As labor progresses, the frequency and intensity of the contraction increases. It is important that uterine contractions are intermittent because a sustained contraction would impede blood flow through the placenta and result in the death of the fetus.

The pain of early uterine activation is due to myometrial hypoxia caused by the compression of blood vessels. Blocking the hypogastric nerves can eliminate this pain. Stretching of the cervix, perineum, and vagina causes the severe pain felt during later stages of uterine activation. The somatic nerves instead of the visceral sensory hypogastric nerve conduct the pain.

Separation and Delivery of the Placenta

Shortly after birth, the uterus drastically decreases in size. This creates a shearing force between the uterine wall and the placenta, separating the placenta from the site of implantation. In species with hemochorial placentation, separation of the placenta from the uterine wall opens blood sinuses resulting in bleeding. This bleeding is usually limited to approximately 350 ml of blood. This limited bleeding occurs in

MAMMALIAN PARTURITION

response to several mechanisms. First, the uterine smooth muscles are arranged in a "figure 8" around the blood vessels as they traverse the uterine wall. Contraction of the uterus following parturition constricts the maternal blood vessels supplying the placenta. Secondly, prostaglandin formed at the placental separation site cause additional spasm of maternal blood vessels.

UTERINE INVOLUTION

<u>Mechanism of Uterine Involution</u>

Uterine involution (phase III) is the reduction of the postpartum human uterus back to the size observed in the nonpregnant state. This is accomplished by sustained contraction of the uterus until uterine size of the nonpregnant state is achieved. Complete involution of the human uterus takes approximately two months. The uterus is less than half of its postpartum weight one week after parturition. By four weeks post partum, the size of the uterus may be the same as that observed before pregnancy. The endometrial surface where the placenta attached undergoes autolysis resulting in **lochia** (a vaginal discharge) for approximately 10 days after parturition. At first, lochia is bloody, but rapidly become serous in appearance. During this time, the epithelial lining of the uterine endometrium is reforming.

FETAL ADJUSTMENTS TO EXTRAUTERINE LIFE

RESPIRATORY ADJUSTMENTS

Lung Liquid Reabsorption during Late Gestation

Lung liquid reabsorption in the human fetus occurs during late gestation. Pulmonary epithelial cells are stimulated by epinephrine and arginine vasopressin resulting in an increase in their intracellular c-AMP (Figure 9.1). The increase in intracellular c-AMP in a pulmonary epithelial cell activates ionic sodium channels on the apical side. The increase in intracellular ionic sodium increases the activity of ionic sodium-potassium-ATPase thus promoting a net increase in the flux of ionic sodium and ionic chlorine from the lung lumen towards the interstitial spaces. This reverses the osmotic gradient across the pulmonary epithelial cell resulting in the reabsorption of lung liquid.

The Onset of Pulmonary Ventilation

The human neonate will usually begin to ventilate within seconds after birth. The promptness in which the neonate initiates breathing indicates that sudden exposure to the outside world causes pulmonary ventilation. The slight asphyxiated state of the fetus and the cooling of the skin following parturition stimulate the respiratory centers of the brain. A neonate that does not immediately start to ventilate becomes hypoxic and hypercapnic. This provides additional stimulation to the respiratory center and usually results in ventilation within an additional few minutes post partum.

Delayed or abnormal pulmonary ventilation at birth can result in hypoxia. Delayed or abnormal pulmonary ventilation in the newborn may be caused by several factors. Intracranial bleeding or brain contusions may physically depress the respiratory centers. Prolonged hypoxia can severely depress the respiratory centers of the brain. Premature separation of the placenta or mechanical compression of the maternal placental blood vessels by excessive uterine contractions can cause neonatal hypoxia. Excessive anesthesia given to the mother during parturition can decrease oxygenation of her blood as well as anesthetize the fetal respiratory centers.

The ability to withstand hypoxia varies considerably between newborns and adults. Cessation of pulmonary ventilation in the adult can result in death within 4 minutes. The fetus can survive hypoxia for 10 minutes to 15 minutes, although lesions in the brain stem become noticeable within 8 minutes to 10 minutes.

Expansion of the Lungs at Birth

The alveoli of the fetal lungs are collapsed by the surface tension of the viscous fluid that fills them. For lung expansion to occur after birth, a negative pressure must be created that exceeds the surface tension within the alveoli. This is accomplished by the first strong inspirations of the newborn. Examination of lung pressure-volume curves of the human newborn (Figure 9.2) shows that the volume of air in the lungs remains almost zero until the negative pressure in the lungs is –40 cm of water. As the negative pressure increases to –60 cm of water, approximately 40 ml of air enters the lungs. To exhale, a considerable positive pressure of 40 cm of water is required to overcome the resistance of the fluid within the bronchioles. The second inspiration is easier requiring a negative pressure of less than –20 cm of water to

FETAL ADJUSTMENTS TO EXTRAUTERINE LIFE

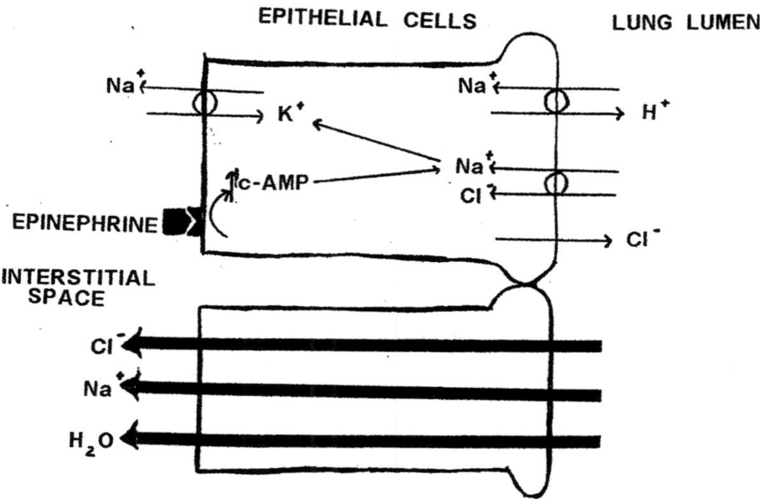

Figure 9.1. The mechanism of liquid reabsorption in the lung of the human fetus during late gestation.

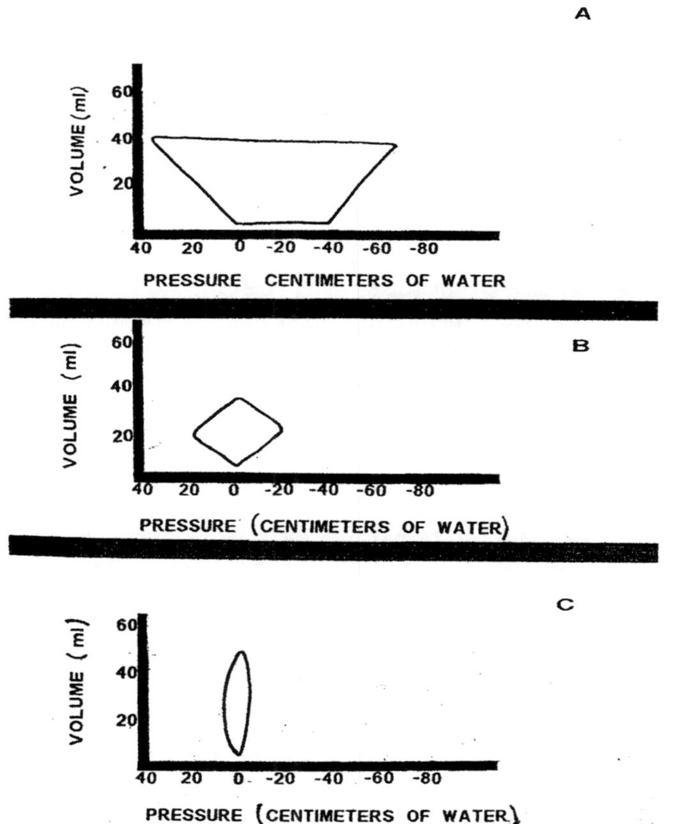

Figure 9.2. Lung compliance curves for human neonates. (A) First breath. (B) Second breath. (C) Forty-five minutes post partum.

overcome the pulmonary resistance. Breathing does not become normal until approximately 40 minutes post partum.

<u>Respiratory Distress in Newborns</u>

One of the most common causes of respiratory distress in human infants is the failure of the lungs to create surfactant. Surfactant is responsible for decreasing the surface tension of alveolar fluids, thereby increasing the compliance of the lungs. Type II alveolar cells do not begin to secrete surfactant until the third trimester. Any premature or full-term newborn without the capability of secreting surfactant has problems expanding their lungs and often develops pulmonary edema.

Hyaline membrane disease is a condition in which the alveoli of the human infant contain large quantities of proteinaceous fluid. Microscopic examination of newborn lungs with hyaline membrane disease shows lung tissue that resembles a hyaline membrane. Premature infants and infants born to diabetic mothers appear to be prone to hyaline membrane disease and often die within hours to several days after birth.

CIRCULATORY ADJUSTMENTS

<u>Changes to Fetal Circulation and Blood Pressure at Birth</u>

Birth causes changes in the systemic vascular resistance and blood pressure in human newborn infants. Removal of the placenta increases blood volume, doubles systemic vascular resistance and increases systemic arterial blood pressure in the newborn (Figure 9.3).

Birth causes changes in pulmonary circulation and blood pressure in human newborns. Vasodilation of pulmonary blood vessels decreases pulmonary vascular resistance resulting in a decrease in pulmonary arterial pressure (Figure 9.4). Vasodilation of pulmonary blood vessels in the newborn is accomplished in two ways. First, fetal pulmonary blood vessels are compressed due to the collapsed lung tissue. Immediately after birth, lung expansion allows fetal pulmonary blood vessels to expand, causing a decrease in pulmonary blood pressure. Secondly, slight fetal hypoxia causes tonic vasoconstriction of pulmonary blood vessels. Oxygenation of the lung tissue at birth causes vasodilation of pulmonary blood vessels. Vasodilation of pulmonary blood vessels causes a decrease in pulmonary blood pressure as well as a decrease in blood pressure in the right chambers of the neonatal heart.

Blood flow is directed to the lungs of the newborn by closure of the foramen ovale. With the left atrial blood pressure being greater (2 mm Hg to 4 mm Hg) than that of the right atrial chamber, blood flow is in the direction of the pressure gradient. Consequently, a small valve in the left atrial chamber closes over the foramen ovale (forming the **fossa ovalis**) preventing blood flow from the left atrial chamber into the right atrial chamber. In most individuals, the valve adheres to the foramen ovale within a few months after birth. Even if permanent adherence does not occur, the slight difference in blood pressure between the left and right atrial chambers keeps the valve closed.

Blood flow is also directed to the lungs of the newborn by closure of the ductus arteriosus. Increased resistance in systemic circulation raises the blood pressure in the aorta, while a decrease in pulmonary vascular resistance decreases the blood pressure in the pulmonary trunk and pulmonary arteries. Within a few hours after birth, the musculature of the ductus arteriosus constricts in response to the elevation in the oxygen content of the blood. Within a week, the constriction is sufficient to prevent blood flow from the

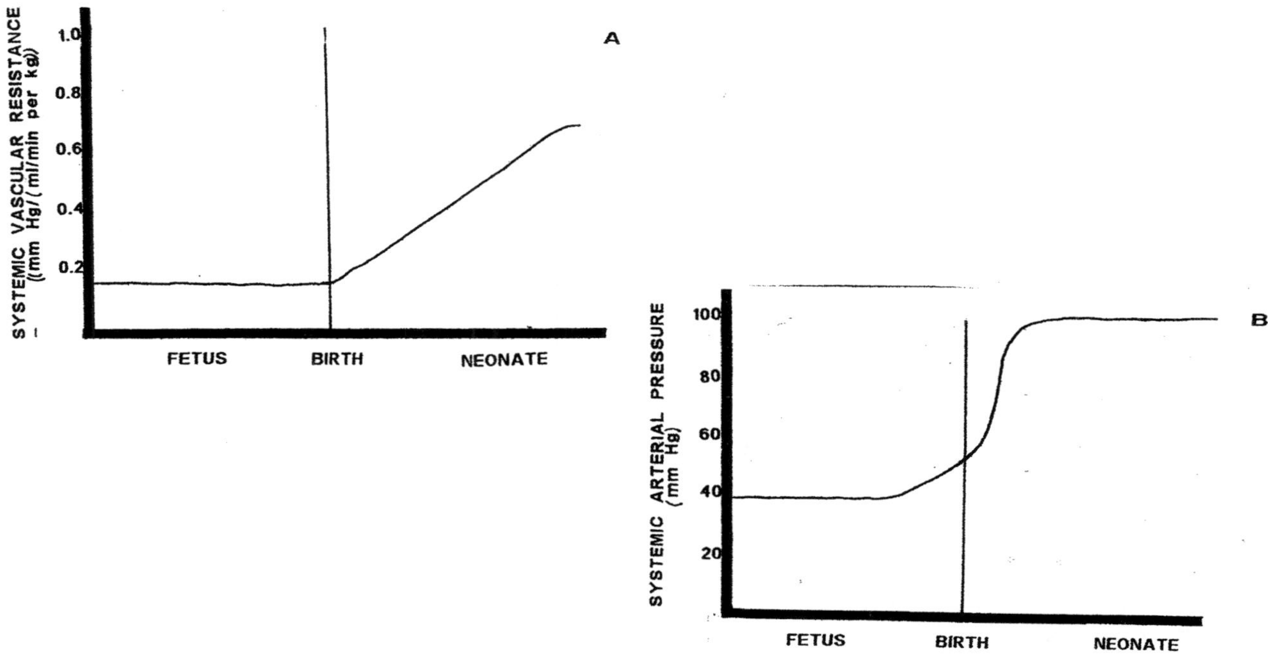

Figure 9.3. Changes in systemic vascular resistance (A) and systemic arterial blood pressure (B) in the human neonate at birth.

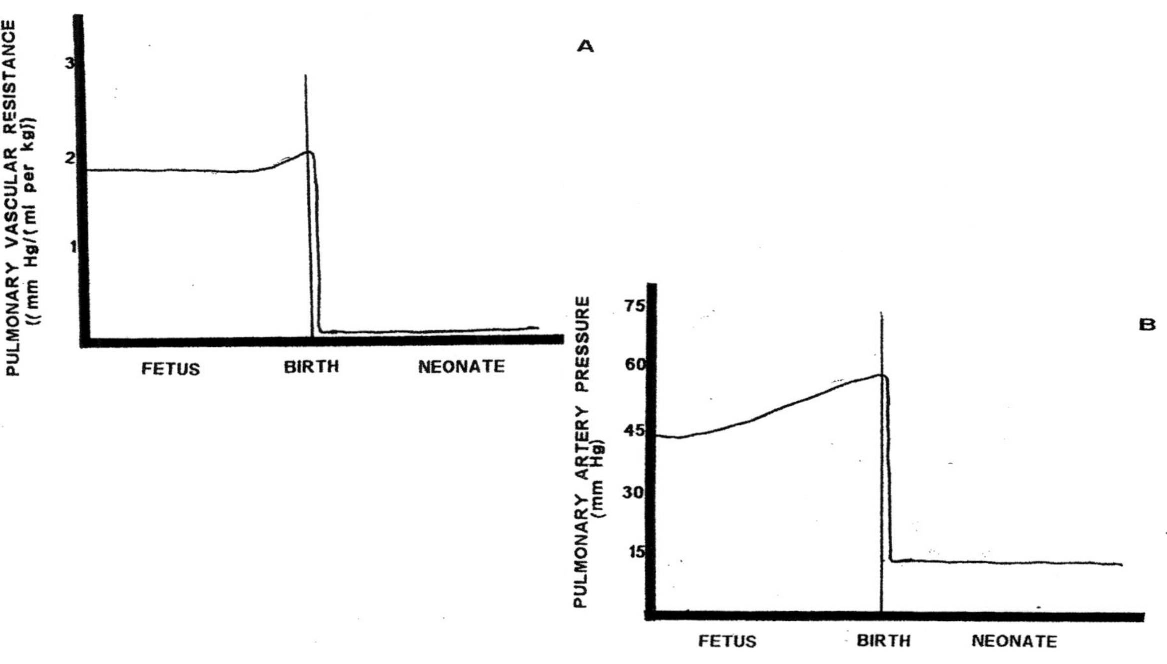

Figure 9.4. Changes in pulmonary vascular resistance (A) and pulmonary arterial blood pressure (B) in the human neonate at birth.

FETAL ADJUSTMENTS TO EXTRAUTERINE LIFE

aorta into the pulmonary vasculature. The ductus arteriosus then becomes occluded by fibrous connective tissue during the next four months forming the **ligamentum arteriosum**.

Blood flow is directed to the liver of the human newborn by closure of the ductus venosus. During fetal life, the portal blood of the fetal abdomen joins the blood flow of the umbilical vein. This blood then passes through the ductus venosus directly into the inferior vena cava. Several hours after birth, the ductus venosus collapses in response to muscular contraction. The portal venous pressure rises forcing the blood to flow through the liver sinuses.

SPECIAL PROBLEMS OF THE NEONATE

The Circulatory System

The blood volume of the newborn human infant is approximately 300 ml. If the newborn is left attached to the placenta for a few minutes or the placental vessels are stripped of blood, an additional 75 ml of blood can enter the systemic circulation of the newborn. Some medical doctors feel that this extra blood volume can cause pulmonary edema and respiratory distress in some newborns. For the first several hours after birth, fluid is lost to the intercellular spaces and the blood volume is returned to basal levels.

The cardiac output of the human newborn is approximately 550 ml/minute. The cardiac output of the neonate in relation to the body weight is twice that observed in the adult. The arterial blood pressure during the first day of life is 70/50, which increases rapidly to 90/60 during the first few months of life. There is a slower rise in blood pressure during the following years until an adult pressure of 115/70 is attained at puberty.

The average erythrocyte count of the human infant is approximately 4 million-cells/cubic millimeter of blood. If the blood is stripped from the placenta at birth, the erythrocyte count can rise an additional 0.5 - 0.75 million-cells/cubic millimeter of blood. During the first 8 weeks to 10 weeks of life, few new erythrocytes are formed since the hypoxic conditions of fetal life is no longer present to stimulate erythrocyte production. The erythrocyte count during this period drops to 3.25 million-cells/cubic millimeter of blood. The increasing activity of the newborn provides the appropriate stimulus for increasing erythrocyte production thus returning the erythrocyte count to within normal parameters within 2 months to 3 months.

Bilirubin is a product of erythrocyte degradation. Bilirubin is formed in the human fetus and transported across the placenta to be excreted by the maternal liver. Immediately after birth, the only way the infant can rid the body of bilirubin is through its own liver. During the first weeks of life, the liver of the newborn functions poorly and is incapable of conjugating significant quantities of bilirubin with glucuronic acid. Therefore, the plasma bilirubin concentration rises by a factor of five during the first three days of life. **Hyperbilirubinemia** is associated with mild jaundice of the infants skin and sclera of the eyes. As the liver becomes functional, blood bilirubin levels return to normal.

Fluid Balance, Acid-Base Balance, and Renal Function

The rate of fluid intake and excretion of the human newborn is seven times greater in relation to the body weight compared to that of the adult. This means that a slight alteration in fluid balance can rapidly cause abnormalities in the newborn. Fetal metabolism of the newborn in relation to the body weight is twice that observed in the adult. This can lead to acidosis in the infant. Functional development of the human newborn kidney is not complete until the first month of life. The newborn kidney concentrates solutes in

FETAL ADJUSTMENTS TO EXTRAUTERINE LIFE

the urine only 1.5 times that observed in blood plasma (adult kidneys can concentrate solutes in the urine 4 to 5 times that observed in blood plasma). Considering the immaturity of the newborn kidney, the marked fluid turnover and the rapid formation of metabolic acids, it is easy to understand that the most important problems in early infancy is acidosis, dehydration, and in some cases overhydration.

<u>Liver Function</u>

Liver metabolic function is deficient in the newborn human infant. The plasma protein concentration of the newborn is 15% to 20% lower than that observed in older children. This indicates that the liver of the newborn infant synthesizes very little protein, Plasma protein concentration can fall so low that the newborn may develop **hypoproteinemic anemia.** Gluconeogenesis is often impaired in the liver of human newborns. Plasma glucose levels of unfed newborns can fall to as low as 4 mg/dl, forcing the infant to depend on fat stores until feeding occurs. The liver of the newborn cannot conjugate bilirubin to glucuronic acid. This allows plasma and tissue bilirubin levels to increase resulting in jaundice during the early days of life. The liver of the newborn is also deficient in producing factors needed for the blood coagulation. This ability of the liver to produced factors needed for blood coagulation does not improve until the infant is several weeks old.

<u>Immunity</u>

The immune system of the human infant is deficient in producing antibodies for the first month of life. The fetus receives maternal antibodies passively across the placenta. The maternal antibodies protect the infant from most major childhood diseases including diphtheria, measles and polio during the first six months of life. Inherited antibodies do not protect the infant against whooping cough. By the end of the first month of life, the infant's plasma antibody levels have decreased to less than half of the levels observed at birth. From this point, the infant's own immune system begins to form antibodies and the gamma globulin concentration of plasma returns to normal levels by 12 months to 20 months of age. Newborn human infants rarely develop allergies. When the neonate's immune system is producing antibodies, allergies can develop often resulting in gastrointestinal abnormalities, eczema, and even anaphylaxis. As children grow older and their immune systems become more competent, these allergies usually disappear.

In mammals with epitheliochorial, synepitheliochorial or endotheliochorial placentation (e.g. swine, cattle, horses, dogs and cats), maternal antibodies do not cross the placenta. Newborn animals must passively acquire maternal antibodies through the milk. In species such as cattle and pigs, it is important that newborns suckle within the first 24 hours to 48 hours of life. After this time, the epithelium of the gut can no longer absorb antibodies.

<u>Neonatal Endocrine Problems</u>

The endocrine system of the human infant is highly developed and endocrine problems and rare. However, endocrine abnormalities do occasionally occur and these are discussed below.

Mothers that carry a female fetus that is exposed to male androgens (e.g. androgen secreting tumors or treatment with androgen hormones) will often give birth to a child with a high degree of masculinization of their sex organs. This condition is called **hermaphroditism.**

Sex hormones secreted by the placenta as well as those secreted by maternal endocrine glands can cause the mammary glands of the neonate to produce milk during the first few days of life. Occasionally, the mammary glands of infants become infected resulting in **infectious mastitis**.

FETAL ADJUSTMENTS TO EXTRAUTERINE LIFE

Infants born to diabetic mothers will often have hypertrophy and hyperfunction of the pancreas. Excess secretion of insulin causes the infants blood glucose concentration to fall lower than 20 gm/dl. Unlike adults, insulin shock and coma in newborns rarely occurs from low levels of glucose.

If the pregnant women is treated with an excess of thyroid hormones or is suffering from hyperthyroidism, the newborn is often born with a temporary hyposecreting thyroid gland. If a thyroidectomy is performed on a woman prior to pregnancy, her pituitary will secrete elevated levels of TSH. This may cause temporary hyperthyroidism in the newborn. Lack of thyroid hormone secretion in the fetus during development causes poor bone development and often results in mental retardation. This condition
is referred to as **cretin dwarfism**.

10

MAMMARY DEVELOPMENT AND LACTATION

ANATOMY OF THE MAMMARY GLAND

<u>Gross Anatomy of the Mammary Gland</u>

The human **mammary gland** (Figure 10.1) is composed of 15 to 20 lobes that are separated by connective tissue. Each lobe is divided into lobules containing alveoli that contain the secreting cells of the gland. Ducts from each alveolus converge into larger ducts that drain the lobule. These large ducts then converge on **secondary ducts** that drain many lobules. Secondary ducts converge into **mammary ducts** that contain a **lactiferous sinus** at their anterior end. Mammary ducts eventually release their contents into the **lactiferous duct** that transports the milk from the lobe to the external environment.

Mammary glands are remarkably similar among different mammalian species (Figure 10.2). The only difference in the gross anatomy of the gland is the removal of the milk gathered in the major ducts. In monotremes, the major ducts draining the mammary gland lead directly to the surface of the skin forming **mammary patches**. Dogs have mammary glands that resemble that of the human. In cattle and goats, the major ducts terminate in a **gland cistern** that is continuous with the **teat cistern**. A **streak canal** drains the teat cistern. The prominence of the gland and the teat cistern varies considerably being quite large in ruminants and small in pigs. In pigs and horses, a single teat empties more than one mammary complex. In this case each teat has two streak canals, one for each gland.

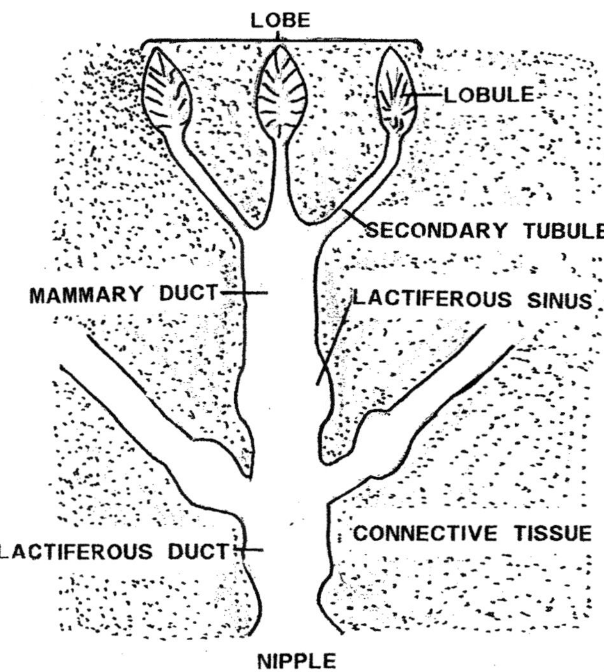

Figure 10.1. Anatomy of the duct system in one lobe of the human mammary gland.

157

MAMMARY DEVELOPMENT AND LACTATION

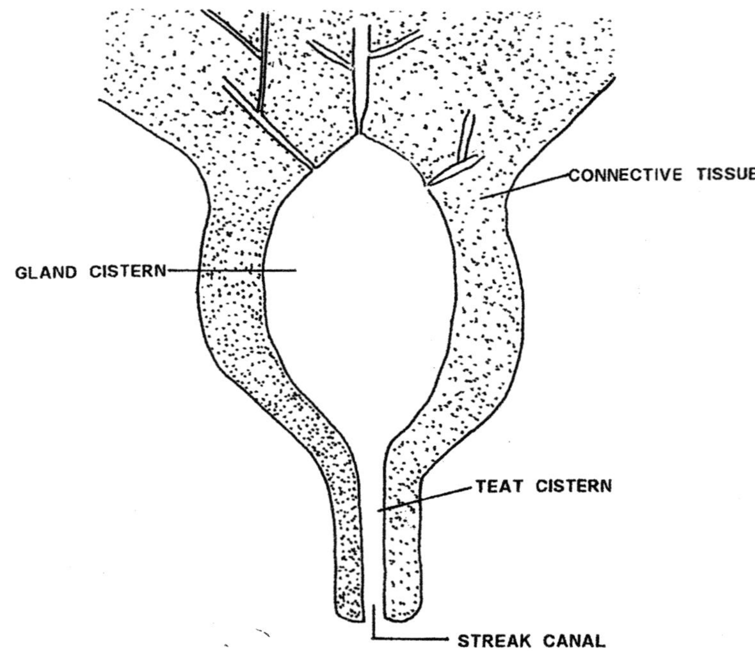

Figure 10.2. Internal anatomy of the cow udder.

<u>Microscopic Anatomy of the Alveolus</u>

The **alveolus** is the basic secretory unit of the mammary gland (Figure 10.3). The alveolus is a small vesicle made of a single layer of epithelial cells that surround a lumen. Each alveolus is surrounded by a basement membrane, a network of capillaries and myoepithelial cells.

Alveoli are arranged in **lobules** and are drained by an **intralobular duct**. These ducts connect to **interlobular ducts** which themselves converge into larger ducts. Lobules are surrounded by connective tissue and are arranged into lobes. The lobes of the mammary gland are also surrounded by connective tissue. Lymphatic capillaries innervate interlobular connective tissue but are not found in intralobular connective tissue.

DEVELOPMENT OF THE MAMMARY GLAND

<u>Development of the Human Mammary Gland from Birth to Puberty</u>

The human mammary gland at birth consists entirely of lactiferous ducts with few alveoli. The human mammary gland remains in this state until the onset of puberty. At puberty, the human female experiences successive cycles of estrogen and progesterone secretion.   Under the influence of estrogen, growth hormone and adrenal steroids the lactiferous ducts start to branch. Solid spheroidal masses composed of polyhedral cells form at the end of the ducts. With the addition of progesterone and prolactin, additional ductal-lobular-alveolar growth occurs. Increased breast size occurs in response to the

MAMMARY DEVELOPMENT AND LACTATION

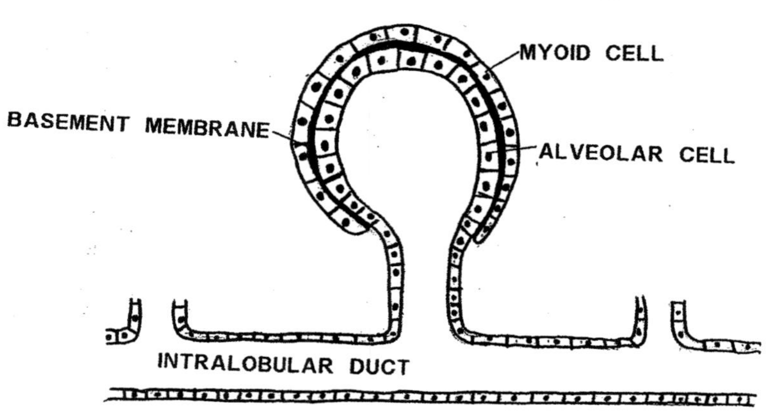

Figure 10.3. Anatomy of a mammary alveolus.

deposition of fat and connective tissues. Many mammalian species differ from humans in that mammary development occurs from midpregnancy to the end of gestation.

Evolution of the Human Mammary Gland

Mammary glands of many female primates develop from midpregnancy to the end of gestation and have a flat appearance. Milk production is dependent on the quantity of glandular tissue in the gland. However, human mammary glands develop at the time of puberty (long before the need to feed an infant!). Compared to the flat mammary gland of most primates, human mammary glands are generally large and rounded. Human mammary glands have an unusually high ratio of fat to glandular tissue.

Propagation within a species depends on individuals having characteristics that allow them to transmit their genes to the next generation. Fast breeders produce more generations per unit of time so their descendants increase at a faster rate compared to slow breeders.   In the history of our species, human males had and continue to have an evolutionary incentive to distinguish mature healthy women from infertile girls. Women had and continue to have an evolutionary incentive to advertise their fertility. Human females tend to be more fertile during their youth. As human females age, they produce fewer children with a higher risk of birth defects. Younger females are usually in better physical condition to care for their offspring and usually their relatives and other offspring also help in caring for the young. An indicator of a human female's age is the shape of her mammary glands. Young females often have large pert mammary glands. As life continues age, gravity and infant suckling cause the females mammary glands to sag. Mammary glands in older females are also less firm.   Human males will favor younger human females with higher fertility. Human males will often favor young females with large pert breasts to mate with. Therefore, attractive benefits in the youth of a human female can outweigh the unattractive costs in older age.

MAMMARY DEVELOPMENT AND LACTATION

Human female mammary glands are a good fitness indicator. When organs grow in pairs, symmetrical development is an indicator of developmental stability. The growth of a symmetrical pair of mammary glands would indicate good developmental stability in the human female. Large mammary glands are also an indicator of a human female's nutritional state. In the Pleistocene, ancestral humans lived on extremely marginal diets interspersed with time periods when food was plentiful. Large breasts would be indicative of the human female's ability to store fat during times of plenty that could be used for reproductive purposes when food was less available.

Development of the Human Mammary Gland during Pregnancy

Further development of the human mammary gland during gestation requires estrogen, progesterone, insulin, prolactin, chorionic somatomammotropin and growth hormone. Growth factors such as epidermal growth factor and transforming growth factor-alpha also appear to play a significant role in regulating mammary growth. The ductular-lobular-alveolar system developed during puberty undergoes further hypertrophy. Prominent lobules form in the mammary gland. The lumens of the alveoli dilate and differentiation of the alveolar cells occurs.

LACTATION

Introduction

There are 2 phases that occur during **lactation** in mammals. **Milk formation** involves the intracellular transport of constituents into the alveolar cells, the synthesis of milk components within alveolar cells, and the discharge of milk into the alveolar lumen. **Milk removal** involves the reflex ejection of milk (milk-letdown) from the alveoli and the removal of milk from the large ducts and cisterns of the mammary gland. These phases of lactation are discussed below.

The Composition of Mammalian Milk

The gross constituents of mammalian milk includes milk fat, lactose, protein, minerals, water, antimicrobial substances, lysozyme, lactoferrin and lactoperoxidase system. The quantity of these gross milk constituents can vary considerably between mammalian species as well as breeds within a given species. The fat concentration in milk can vary from a trace to 500 grams/leter. Lactose concentration in mammalian milk can vary from a trace to 100 grams/leter. The protein concentration of mammalian milk can vary from 10 grams/leter to 200 grams/leter.

Milk fat has a melting point that is lower than body temperature. Milk fat is dispersed in milk as droplets surrounded by a membrane of phospholipid-protein complexes. Mammalian milk fat is composed of a combination of triglycerides, diglycerides, monoglycerides, free fatty acids and sterol. The type of fatty acids in milk varies considerably between mammalian species. Human milk contains triglycerides with palmitic and oleic being the most common fatty acids. The concentration of fat in human milk ranges from 3% to 5%.

Carbohydrates can vary in the milk of different mammalian species. The principal sugar found in the milk of marsupials is galactose. Lactose is the chief sugar found in the milk of eutherian mammals. Lesser carbohydrates found in the milk of eutherian mammals include glucose, galactose, fucose and N-acetylglucosamine. The concentration of lactose in human milk is approximately 7%.

MAMMARY DEVELOPMENT AND LACTATION

Several proteins are found in mammalian milk that can vary in concentration between species. Milk proteins include alpha-casein, lactalbumin, lactoferrin, immunoglobulin A, lysozyme and albumin. Total protein concentration in human milk is approximately 1%.

Mammalian milk contains numerous ions and several growth factors. These ions include sodium, potassium, calcium, magnesium, phosphorus and chlorine. The concentration of these ions in human milk is approximately 0.2%. Epidermal growth factor, transforming growth factor, somatostatin as well as IGF-I and IGF-II can also be found in the milk of mammals.

In some mammalian species milk can serve as a vehicle for the transfer of antibodies from the mother to the infant. Mammals with hemochorial placentas (e.g. humans and monkeys) and hemoendothelial placentas (e.g. rabbit) primarily transfer maternal antibodies directly into the blood of the fetus. Mammals with epitheliochorial placentas (e.g. cattle and swine) and endotheliochorial placentas (e.g. dogs and cats) secrete **colostrum** for the newborn to ingest immediately after parturition. Colostrum has a higher concentration of protein and sodium chloride, and is lower in lactose and potassium when compared to normal milk. **Immunoglobulin A** is an important protein component of colostrum that provides **passive immunity** to the newborn. Immunoglobulin A is produced by plasma cells located in the mammary gland. These plasma cells are derived from sensitized B-lymphocytes that migrated from Peyer's patches [Johann Peyer, Swiss anatomist, 1653 – 1712] of the intestine. Estrogen, progesterone and prolactin regulate the migration of these plasma cells in mice. Plasma cells in the mammary gland can secrete antibodies against organisms ingested by the mother. During the secretion of milk, **secretory immunoglobulin A** is formed (immunoglobulin A forms molecular dimers with glycoprotein attachments). Secretory immunoglobulin A has an increased resistance to exposure to enzymes and pH that would be found in the gut of the neonate.

The composition of mammalian milk can vary in relation to environmental conditions as well as the nursing frequency. In certain Arctic, aquatic and desert mammalian species fat is the main constituent of milk. A high concentration of milk fat aids the young in offsetting heat loss to the external environment in Arctic and aquatic mammals, A high milk fat content aids in maternal water conservation in dessert mammals. Mammalian species that tend to nurse frequently have milk with lower concentration of nutrients compared to those mammalian species that nurse less frequently.

<u>Milk Formation</u>

Milk formation in the alveolus of a human mammary gland involves a minimum of 4 transcellular pathways. The secretion of milk from the alveolar cells into the alveolar lumen occurs continuously, but milk does not enter the mammary duct system. The absorption of water into the alveolar cell involves water moving across the plasma membrane by osmosis. The osmosis of water into the alveolar cell occurs in response to the solute concentration gradient generated by intracellular lactose. Monovalent cations are attracted into the alveolar cell by electrical attraction. Immunoglobulins are absorbed into the alveolar cell by receptor mediated endocytosis. Immunoglobulin A is transported to the Golgi apparatus and eventually to the apical membrane to be secreted in the milk. The synthesis of milk lipids occurs in the cytoplasm and Golgi apparatus of the alveolar cell. Milk lipid droplets coalesce into large fat droplets within the alveolar cell that eventually attach to the apical membrane of the cell. The plasma membrane of the alveolar cell pinches off releasing membrane enclosed fat droplets into the alveolar lumen. Milk proteins are produced within the alveolar cells. These proteins are vacuolized by the Golgi apparatus and secreted into the alveolar lumen by exocytosis. Lactose is produced in the Golgi apparatus of the alveolar cell. Lactose combines with alpha-lactalbumin (whey) and the enzyme galactosyltransferase to form lactose synthetase. The sugar then passes across the cell membrane of the alveolar cells with the protein granules into the alveolar lumens. Osmotically active lactose is primarily responsible for determining the liquid volume of milk. Proteins, immunoglobulin A and leukocytes found in the maternal plasma can enter the alveolar

MAMMARY DEVELOPMENT AND LACTATION

lumen by moving between alveolar cells that loose their tight junctions.

Prolactin and other synergistic hormones primarily regulate the quantity and content of mammalian milk. Prolactin stimulates the synthesis of B-lactoglobulin and casein as well as stimulating milk fat synthesis in human mammary glands that have previously been primed by insulin and cortisol. Prolactin increases the half-life of casein m-RNA by a factor of eight. Prolactin appears to be involved in the transport of ionic sodium into mammary tissue.

<u>The Initiation and Maintenance of Lactation</u>

The onset of mammary secretion varies between mammalian species. The initiation of lactation in some mammalian species occurs during the last third of gestation, whereas the initiation of lactation in other mammalian species occurs at the time of parturition. In rodents, a decrease in plasma progesterone approximately 30 hours prior to parturition allows prolactin, placental lactogen and adrenal steroids to assume their milk-producing role. In other mammalian species loss of the placenta following parturition decreases the plasma levels of estrogen and progesterone. The decrease in plasma levels of estrogen and progesterone allows prolactin to conduct its milk-producing role in conjunction with growth hormone, cortisol, parathyroid hormone and insulin.

The maintenance of lactation requires the presence of several hormones. Growth hormone is required for the maintenance of lactation in ruminant species whereas prolactin is required in nonruminant species. The concentration of plasma prolactin through human pregnancy increases from approximately 20 ng/ml to 200 ng/ml by the end of gestation. Plasma prolactin levels remains elevated for approximately 4 weeks to 6 weeks in the post-partum female, but then falls to levels observed in the nonpregnant state. A 10 to 20 fold increase in plasma prolactin concentration occurs each time a human female nurses (Figure 10.4). This increased prolactin acts on the mammary gland to keep it functional for subsequent nursing periods. Milk production in the human mammary gland can occur for several years if the child continues to suckle at frequent intervals; however, the rate and amount of milk produced decreases considerably within 7 to 9 months post partum. The hypothalamus secretes two hormones that alter the synthesis and secretion of prolactin from the anterior pituitary gland. Prolactin inhibiting hormone (dopamine) tonically inhibits prolactin production and secretion, whereas prolactin-releasing hormone can intermittently increase the production and secretion of prolactin. Suppression of the female ovarian cycle can occur during nursing. In approximately 50% of nursing women, the ovarian cycle does not resume until a few weeks after the cessation of nursing. It appears that the same bioelectrical information that causes prolactin secretion during suckling also inhibits the secretion of gonadotropic releasing hormone by the hypothalamus. This in turn inhibits the synthesis and secretion of gonadotropic hormones from the anterior pituitary. After several months of lactation, the pituitary gland in 50% of lactating women begins to secrete FSH and LH to reinstate the menstrual cycle.

<u>Milk Ejection from the Mammary Gland</u>

Milk is continuously produced by the alveoli of the mammary glands during the period of lactation. Milk does not flow easily from the alveoli into the duct system of the mammary gland, but must be ejected into the ducts before the newborn can obtain it. This process involves a neuroendocrine reflex (Figure 10.5). When the newborn first suckles, it virtually does not receive any milk. Bioelectrical sensory information is transmitted from the nipples to the spinal cord and eventually to the hypothalamus. Bioelectrical stimulation of the hypothalamus causes the synthesis and release of oxytocin from the paraventricular nucleus. Oxytocin is then transported to the neurohypophysis via the hypothalamic-hypophyseal tract. The neurohypophysis then secretes oxytocin into the blood where it travels to and stimulates the mammary gland. Oxytocin causes the myoepithelial cells of the mammary gland to contract, ejecting milk into the duct system. Suckling by the newborn then removes the milk from the mammary gland.

Milk ejection can be inhibited by many factors. Many psychogenic factors and generalized sympathetic stimulation can inhibit oxytocin secretion and consequently lower milk ejection. For this reason it is important that the mother and offspring be undisturbed if successful nursing is to occur.

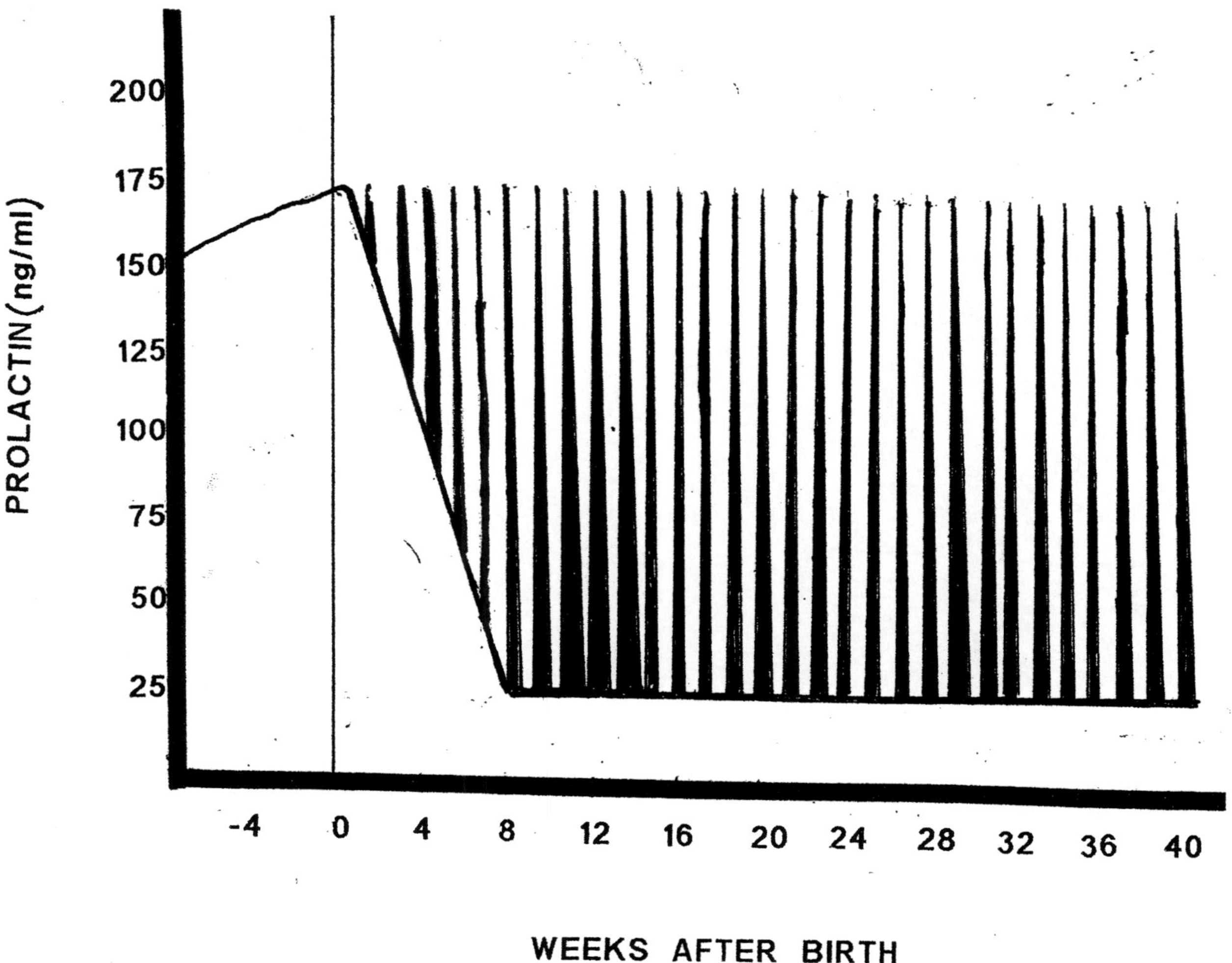

Figure 10.4. The secretion of prolactin during nursing in the human.

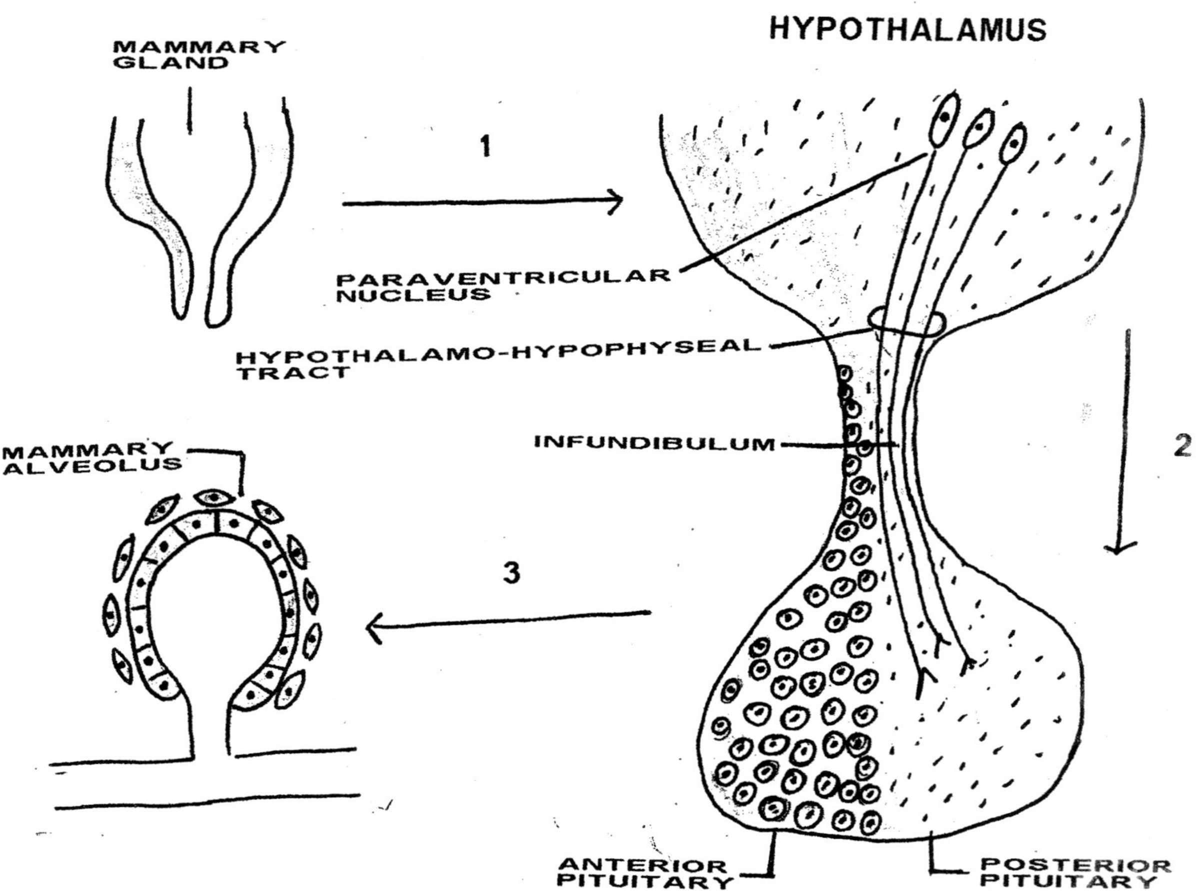

Figure 10.5. A neuroendocrine reflex of milk-let down. (1) Suckling stimulates the mammary receptors sending sensory information to the paraventricular nucleus of the hypothalamus. (2) Paraventricular nucleus produces and secretes oxytocin that is transported via the hypothalamic-hypophyseal tract to the posterior pituitary. (3) Posterior pituitary release oxytocin into the blood. Oxytocin stimulates myoid cells around mammary alveolus to contract resulting in milk ejection

SOURCES

SOURCES

CHAPTER 1

Clark, W. 1999. **A Means to an End. The Biological Basis of Aging and Death**.
  Oxford University Press.
Margulis, L. 1998. **Symbiotic Planet. A New Look at Evolution.** Basic Books.
Margulis, L. and Sagan, D. 1997. **What is Sex.** Simon and Schuster.
Margulis, L. and Sagan, D. 1995. **What is Life.** Simon and Schuster.
Margulis, L. and Sagan, D. 1986. **Origins of Sex. Three Billion Years of Genetic Recombination.**
  Yale University Press.
Michod, R. 1995. **Eros and Evolution. A Natural Philosophy of Sex.** Addison Wesley.
Nester, E., Roberts, C., McCarthy, B. and Persall, N. 1973. **Microbiology. Molecules, Microbes and
  Man.** Holt, Rinehart and Winston.
Smith, J.M. and Szathmary, E. 1999. **The Origins of Life. From the Birth of Life to the Origin of
  Language.** Oxford University Press.
Smith, J. M. and Szathmary, E. 1995. **The Major Transitions in Evolution.** Oxford University Press.

CHAPTER 2

Aron, D., J. Findling and J. Tyrell. 2001. Hypothalamus and pituitary. In: **Basic and Clinical
  Endocrinology.** 6th edition. (F. Greenspan and D. Gardner, eds.), McGraw Hill.
Baulieu, E. 1990. Receptor-associated nuclear proteins and earliest effect of steroid hormones. In:
  Neuroendocrine Regulation of Reproduction. (S. Yen and W. Vale, eds.). Serono
Bohnet, H. and Friesen, H. 1976. Effect of prolactin and growth hormone on prolactin and LH receptors
  in the dwarf mouse. **J. Reprod. Fert. 43:307.**
Bolander, F. 1989. **Molecular Endocrinology.** Academic Press.
Bremmer, N., Batatel, C., McLane, R., Matsumoto, A., Tenover, J. and Soules, M. 1991. Central
  control of inhibin secretion in humans. In: **Neuroendocrine Regulation of Reproduction.**
  (S. Yen and W. Vale, eds.). Serono.
Bronson, F. 1983. Hormonal response and primer pheromones. In: **Pheromones and Reproduction in
  Mammals.** (J. Vandenburgh, ed.) Academic Press.
Chemes, H., Rivarola, M. and Berganda, C. 1976. Effect of hCG on the interstitial cells and androgen
  production in the immature rat testes. **J. Reprod. Fert. 46:279.**
Chemes, H., Gottlieb, S. and Pasqualini, T. 1985. Response to acute hCG stimulation and steroidgenic
  potential of Leydig cell fibroblast precursers in the human. **J. Androl, 6:102.**
Clark, I., Phillips, D., Mercer, J. and Robertson, D. 1991. Mechanisms involved in LH surge. In:
  Neuroendocrine Regulation of Reproduction. (S. Yen and W. Vale, eds.). Serono.
Colenbrander, B., deJong, F. and Wensing, C. 1978. Changes in serum testosterone concentration in the
  male pig during development. **J. Reprod. Fert. 53:377.**
Colenbrander, B., van de Wiel, D., van Rossum-Kok, C. and Wensing, C. 1982. Changes in FSH
  concentrations in the pig during development. **Biol. Reprod. 26:105.**
Curot, M. 1967. Endocrine control of the supporting and the germ cells of the impubertal testes.
  J. Reprod. Fert. (Suppl.) 2:89.
Danchin, E. and Dubois, M. 1982. Immunocytochemical study of the chronology of pituitary cytogenesis
  in the domestic pig (Sus scofa) with special reference to the functioning of the hypothalamo-
  pituitary gonadal axis. **Reprod. Nutr. Develop. 22:135.**
Elsaesser, F., Parvizi, N., and Ellendorff, F. 1978. Steroid feedback in luteinizing hormone and secretion
  during sexual maturation in the pig. **J. Endocr. 78:329.**

SOURCES

Fritz, I. 1978. Sites of action of androgens and follicle stimulating hormone on the cells of the seminiferous tubule. In: **Biochemical Action of Hormones**. (B. Lutwak, ed). Academic Press.

Ganong, W. 1991. Role of the nervous system in reproduction. In: **Reproduction in Domestic Animals**. 4 ed. (P. T. Cupps, ed.) Academic Press.

Gordon, J. and L. Speroff. 2002. **Handbook of Clinical Gynecological Endocrinology and Infertility**. Lippencott, Williams and Wilkins.

Gray R., Day, B., Lasley, J. and Tribble, L. 1971. Testosterone levels in boars of various ages. **J. Anim. Sci. 33: 124.**

Grumbach, M. 1980. The neuroendocrinology of puberty. In: **Neuroendocrinology.** (D. T. Krieger and J. Hughes, eds.). Sinauer.

Hendricks, D. 1991. Biochemistry and physiology of gonadal hormones. In: **Reproduction in Domestic Animals.** 4 ed. (P. T. Cupps, ed.). Academic Press.

Izard, M. 1983. Pheromones and reproduction in domestic animals. In: **Pheromones and Reproduction in Mammals.** (J. Vandenburgh, ed.). Academic Press.

Karsch, F. 1984. The hypothalamus and the anterior pituitary. In: **Reproduction in Mammals. Hormonal Control of Reproduction.** 2 ed. (C. R. Austin and R. V. Short , eds.). Cambridge University Press.

Klindt., J. and Stone, R. 1984. Porcine growth hormone and prolactin. Concentrations in the fetus and secretory pattern in the growing pig. **Growth 48:1**.

Kosco, M. and J. Hill. 2003. **Principles of Mammalian Physiology**. Kosco Press.

Ledger, W. 2001. Inhibin and activin in the male. In: **Inhibin, Activin and Follistatin in Human Reproductive Physiology.** (S. Muttukrishna amd W. Ledger, eds.) Imperial College Press.

Lehninger, A. 1975. **Biochemistry.** 2 ed. Worth Publishers

Lincoln, D. 1984. The posterior pituitary. In: **Reproduction in Mammals. Hormonal Control of Reproduction.** 2 ed. (C. R. Austin and R. V. Short, eds.). Cambridge University Press.

Liwiska, J. 1975. Development of the adenohypophysis in the pig. **Folia. Morphol. 34:211.**

Padmanablan, V. and C. West. 2001. Endocrine, autocrine and paracrine actions of inhibin, activan, follistatin on follicle stimulating hormone. In: **Inhibin, Activin and Follistatin in Human Reproductive Physiology.** (S. Muttukrishna amd W. Ledger, eds.) Imperial College Press.

Pfaff, D., Kow, L., Bergen, H. and Schwanzel-Fukuda, M. 1990. Luteinizing hormone releasing hormone neurons as they develop and effect other neurons. In: **Neuroendocrine Regulation of Reproduction**. (S. Yen and W. Vale, eds). Serono.

Porterfield, S. 2001. **Endocrine Physiology**. 2nd edition. Mosby.

Reiter, E. and Grumbach, M. 1982. Neuroendocrine control mechanisms and the onset of puberty. **Annu. Review Physiol. 44:595.**

Skinner, M. 1990. Cell-cell interaction in the testis which regulate gonadal function. In: **Neuroendocrine Regulation of Reproduction.** (S. Yen and W. Vale, eds.) Serono.

Ward, D., Bousfield, G., and Moore, K. 1991. Gonadotropins. In: **Reproduction in Domestic Animals.** 4 ed. (P. T. Cupps, ed.). Academic Press.

CHAPTER 3

Abdel-Raouf, M. 1960. The postnatal development of the reproductive organs of the bull with special reference to puberty. **Acta Endocrinologica 34 (Suppl. 49): 1-109**.

Abou-Haila, A. and D. Tulsiani. 2003. The sperm acrosome: formation and contents. In: **Introduction to Mammalian Reproductive Physiology.** (D. Tulsiani, ed.). Kluwer Academic Publishers.

SOURCES

Aoki, B. 1970. Hormonal control of Leydig cell differentiation. **Protoplasma 71:209.**

Aoki, B. and Fawcett, D. 1978. Is there a local feedback from the seminiferous tubules affecting the activity of Leydig cells? **Biol. Reprod. 19:144.**

Aquas, A. 1981. Direct contact between actin-like microfilaments and organelles in pig Leydig cells. **J. Ultrastructural Res. 74:175.**

Attal, J. 1969. Levels of testosterone, androstenedione, estrone and estradiol 17B in the testes of fetal sheep. **Endocrinology 85:280.**

Baker, R. 1996. **Sperm Wars. The Science of Sex.** Basic Books.

Baker, R. and M. Bellis. 1995. **Human Sperm Competition.** Chapman and Hall.

Belleve, A. and Feig, L. 1984. Cell proliferation in the mammalian testis: Biology of seminiferous growth factor (SGF). **Recent Progress in Hormone Research 40:531.**

Belt, W. 1967. Fine structure of the interstitial cell of the boar. **Anat. Rec. 158:333.**

Beyler, S. and L. Zanefeld. 1982. The male accessory sex glands. In: **Biochemistry of Mammalian Reproduction.** John Wiley and Sons.

Black, V. and Christenson, K. 1969. Differentiation of interstitial cells and Sertoli cells in the fetal quinea pig. **Am. J. Anat. 124: 211.**

Bols, N., Bowen, B. and Byrne, P. 1984. Rat Sertoli cells in culture undergo metabolic co-operation with each other and with fibroblasts. **J. Reprod, Fert. 70:191.**

Brandes, D. 1974. Fine structure and cytochemistry of male accessory organs. In: **Male Accessory Organs. Structure and Function in Mammals.** (D. Brandes, ed.) Academic Press.

Bronson, F. 1989. **Mammalian Reproductive Biology.** University of Chicago Press.

Bruhn, T., Ponzilius, K. and Ellendorff, F. 1984. Ontogeny of hypothalamic control of gonadotropin secretion in the fetal pig. **10th International Congress on Animal Reproduction and Artificial Insemination.** June 10-14 Urbana-Champaign, U.S.A., 4:V10.

Burgos, M. 1974. Biochemical and functional properties related to metabolism. In:**Male Accessory Organs. Structure and Function in Mammals.** (D. Brandes, ed.) Academic Press.

Calamita, G., A. Mazzone, Y. Cho. G. Valenti and M. Svelto. 2001. Expression and localization of the aquaporin-8 water channel in rat testis. **Biol. Reprod. 64:1660.**

Chemes, H., Rivarola, M. and Berganda, C. 1976. Effect of hCG on the interstitial cells and androgen production in the immature rat testes. **J. Reprod. Fert. 46:279.**

Chemes, H., Gottlieb, S. and Pasqualini, T. 1985. Response to acute hCG stimulation and steroidgenic potential of Leydig cell fibroblast precursers in the human. **J. Androl, 6:102.**

Chiarini-Garcia, H. and L. Russell. 2001. High-resolution light microscopy characterization of mouse spermatogonia. **Biol Reprod 65:1170.**

Christenson, A. and Gillium, S. 1969. The correlation of fine structure and function in steroid secreting cells, with emphasis on those of the gonad. In: **The Gonads.** (K. W. McKerns, ed.). North Holland Publishing Company.

Colenbrander, B., deJong, F. and Wensing, C. 1978. Changes in serum testosterone concentration in the male pig during development. **J. Reprod. Fert. 53:377.**

Colenbrander, B., van de Wiel, D., van Rossum-Kok, C. and Wensing, C. 1982. Changes in FSH concentrations in the pig during development. **Biol. Reprod. 26:105.**

Cornwall, G. 2003. Ductus epididymus. In: **Introduction to Mammalian Reproductive Physiology.** (D. Tulsiani, ed.). Kluwer Academic Publishers.

Crabo, B. 1965. Studies on the composition and epididymal content of bulls and boars. **Acta Veterinaria Scandinavica 6: Suppl. 5.**

Curot, M. 1967. Endocrine control of the supporting and the germ cells of the impubertal testes. **J. Reprod. Fert. (Suppl.) 2:89.**

Dacheux, J. and I. Dacheux. 2002. Protein secretion in the epididymus. In: **The Epididymis. From Molecules to Clinical Practice.** (B. Robaire and B. Hinton, eds.). Kluwer Academic/Plenum Publishers.

SOURCES

Dacheux, F. 1978. Ultrastructural localization of gonadotropic hormones in the porcine pituitary using immunoperoxidase technique. **Cell Tiss. Res. 191:219.**

Dacheux, F. and Martinat, N. 1983. Immunocytochemical localization of LH, FSH, and TSH in the fetal porcine pituitary. **Cell Tiss. Res. 228:277.**

Danchin, E. and Dubois, M. 1982. Immunocytochemical study of the chronology of pituitary cytogenesis in the domestic pig (Sus scofa) with special reference to the functioning of the hypothalamo-pituitary gonadal axis. **Reprod. Nutr. Develop. 22:135.**

deKretser, D. 1984. The testis. In: **Reproduction in Mammals. Hormonal Control of Reproduction.** 2 ed. (C. R. Austin and R. V. Short, eds.). Cambridge University Press.

de Rooij, D. 2001. Proliferation and differentiation of spermatogonial stem cells. **Reproduction 121: 347.**

Dierichs, R., Wrobel, K. and Schilling, E. 1973. Light-und elektronen mikroskopische Inter-suchungen an de Leydig zellan des schweines wahrend der postnatalen Entwicklung. **Z. Zellforsch 143:207.**

Eberhard, W. 1996. **Female Control: Sexual Selection by Female Choice.** Princeton University Press.

Eberhard, W. 1985. **Sexual Selection and Animal Genitalia.** Harvard University Press.

Elsaesser, F., Parvizi, N., and Ellendorff, F. 1978. Steroid feedback in luteinizing hormone and secretion during sexual maturation in the pig. **J. Endocr. 78:329.**

Finnson, D., J. Dufresne and M. Gregory. 2002. Cellular interactions and the blood epididymal barrier. In: **The Epididymis. From Molecules to Clinical Practice.** (B. Robaire and B. Hinton, eds.). Kluwer Academic/Plenum Publishers.

Ford, J. 1983. Serum estrogen concentrations during postnatal development in the male pig. **Proc. Soc. Exptl. Biol. Med. 174:160.**

Ford, J., Christenson, R., and Maurer, R. 1980. Serum testosterone concentrations in embryonic and fetal pigs during sexual differentiation. **Biol. Reprod. 23:583.**

Fritz, I. 1978. Sites of action of androgens and follicle stimulating hormone on the cells of the seminiferous tubule. In: **Biochemical Action of Hormones.** (B. Lutwak, ed). Academic Press.

Gray R., Day, B., Lasley, J. and Tribble, L. 1971. Testosterone levels in boars of various ages. **J. Anim. Sci. 33: 124.**

Gordon, J. and L. Speroff. 2002. **Handbook of Clinical Gynecologic Endocrinology and Infertility.** Lippincott, Williams and Wilkins.

Guraya, S. 2001. **Comparative Cellular and Molecular Biology of the Testis of Vertebrates.** Science Publishers, Inc.

Guraya, S. 2000. Cellular and molecular biology of capacitation and acrosome reaction in spermatozoa. **Int. Rev. Cyt. 199:1.**

Guraya, S. 1999. **Comparative Testicular Biology in Animals.** Science Publishers Inc.

Hannson, V., French, F., Weddington, S., Neyfey, S. and Ritzen, G. 1974. FSH stimulation of androgen testicular binding protein (ABP): Androgen priming increase the response to FSH. In: **Hormone Binding and Target Cell Activation in the Testis.** (M. Dufau and A. Means, eds.). Plenum Press.

Hannson, V., Weddington, S., Naess, O. and Attramadal, A. 1975. Testicular androgen binding protein (ABP)—A parameter of Sertoli cell secretory function. In: **Hormonal Control of Spermatogenesis.** (F. French, V. Hannson, E. Ritzen, and S. Neyfey, eds.). Plenum Press.

Hassold, T. 1998. Nondisjunction in the human male. In: **Meiosis and Gametogenesis.** (M. Handel, ed.) Academic Press.

Herrera, J., Sosa, M., Ayala, F., Galina, C., Javrequi, P. and Bermudez, J. 1983. Morphophysiological correlation of boar Leydig cell development during the postnatal stage. **Cornell Vet. 73:67.**

Hermo, L. and B. Robaire. 2002. Epididymal cell types and their functions. In: **The Epididymis. From Molecules to Clinical Practice.** (B. Robaire and B. Hinton, eds.). Kluwer Academic/Plenum Publishers.

Hess, R. 2002. The efferent ducts. In: **The Epididymis. From Molecules to Clinical Practice.** (B. Robaire and B. Hinton, eds.). Kluwer Academic/Plenum Publishers.

SOURCES

Hullinger, R. and Wensing C. 1985. Testicular organogenesis in the fetal calf. Interstitial (Leydig) development. **Acta Anat. 121:99**.

Hyman, L. 1942. **Comparative Vertebrate Anatomy.** University of Chicago Press.

Isaksson, O., Eden, S. and Jansson, J. 1985. Mode of action of pituitary growth hormone on target cells. **Ann. Rev. Physiol. 47:483**.

Izard, M. 1983. Pheromones and reproduction in domestic animals. In: **Pheromones and Reproduction in Mammals.** (J. Vandenburgh, ed.). Academic Press

Johnson, L. 1991. Spermatogenesis. In: **Reproduction of Domestic Animals.** 4th ed. (P. T. Cupps, ed.). Academic Press.

Junquiera, L., Carneiro, J. and Kelley, R. 1998. **Basic Histology.** 9th ed. Appleton and Lange.

Kosco, M. and J. Hill. 2003. **Principles of Mammalian Physiology**. Kosco Press.

Kosco, M., Crabo, B. and Loseth, K. 1989. Seminiferous tubule response in relation to compensatory testicular growth after neonatal hemicastration. **J. Reprod. Fert. 87:1**.

Kosco, M., Crabo, B. and Loseth, K. 1989. Interstitial response in relation to compensatory testicular growth after neonatal hemicastration in boars. **J. Reprod. Fert. 87:13**.

Kosco, M., Crabo, B., Bolt, D. and Loseth, K. 1987. Endocrine response in relation to compensatory testicular growth after neonatal hemicastration in the boar. **Biol. Reprod. 36:1177**.

Kuntz, A. and Morris, E. 1946. Components and distribution of the spermatic nerves and the nerves of the vas deferens. **J. Comp. Neurol. 85:33**.

Mann, T. 1964. **The Biochemistry of Semen and the Male Reproductive Tract**. Methuen.

Margulis, L. and D. Sagan. 1991. **Mystery Dance. The Evolution of Human Sexuality**. Summit Books.

Miller, G. 2000. **The Mating Mind**. Doubleday.

Moreno, R., J. Ramalho-Santos, P. Sutovsky, E. Chan and G. Schatten. 2000. Vesiculae traffic and Golgi apparatus dynamics during mammalian spermatogenesis. **Biol. Reprod. 63:89**.

Narbaitz, R. 1974. Embryology, anatomy and histology of the male accessory glands. In: **Male Accessory Organs. Structure and Function in Mammals.** (D. Brandes, ed.) Academic Press.

Neelankanta, R., L. Dettin, and M. Dym. 2003. Mammalian testes:structure and function. In: **Introduction to Mammalian Reproductive Physiology.** (D. Tulsiani, ed.). Kluwer Academic Publishers.

Rodriguez-Rigau, L. and E. Steinberger. 1982. The testis and spermatogenesis. In: **Biochemistry of Mammalian Reproduction.** (L. Zanefeld and R. Chatterton, eds.) John Wiley and Sons.

Rodriguez, C. and B. Hinton. 2003. The testicular and epididymal luminal fluid microenvironment. In: **Introduction to Mammalian Reproductive Physiology**. (D. Tulsiani, ed.). Kluwer Academic Publishers.

Setchell, B. 1991. Male reproductive organs and semen. In: **Reproduction in Domestic Animals.** 4 ed. (P. T. Cupps, ed.) Academic Press.

Setchell, B. 1982. Spermatogenesis and spermatozoa. In: **Reproduction in Mammals. Germ Cell and Fertilization.** (C. R. Austin and R. V. Short, eds.). Cambridge University Press.

Setchell, B. 1978. **The Mammalian Testis**. Cornell University Press.

Vandenburgh, J. 1983. Pheromonal regulation of puberty. In: **Pheromones and Reproduction in Mammals.** (J. Vandenburgh, ed.) Academic Press.

Wheater, P., Burkitt, H. and Daniel, V. 1979. **Functional Histology.** Churchhill and Livingston.

Zanefeld, L. 1982. The epididymis. In: **Biochemistry of Mammalian Reproduction**. (L. Zanefeld and R. Chattterton, eds.) John Wiley and Sons.

SOURCES

CHAPTER 4

Albertini, D. 2002. The structural basis of oocyte-granulosa cell communication. In: **The Future of the Oocyte. Basic and Clinical Application.** (J. Eppig, C. Hegele-Hartung and M. Lessl, eds.) Springer-Verlag.

Anesetti, G., P. Lombide, H. D' Albora and S. Ojeda. 2001. Intrinsic neurons in the human ovary. **Cell Tissue Res. 306:231.**

Angler, N. 1999. **Women: An Intimate Geography.** Houghten and Mifflen.

Austin, C. 1982. The egg. In: **Reproduction in Mammals. Germ Cells and Fertilization.** 2nd ed. (C. R. Austin and R. V. Short, eds.). Cambridge University Press.

Baird, D. 1984. The ovary. In: **Reproduction in Mammals. Hormonal Control of Reproduction.** 2nd ed. (C. R. Austin and R. V. Short, eds.). Cambridge University Press.

Baker, T. 1982. Oogenesis and ovulation. In: **Reproduction in Mammals. Germ Cells and Fertilization.** 2nd ed. (C. R. Austin and R. V. Short, eds.). Cambridge University Press.

Bolander, F. 1989. **Molecular Endocrinology.** Academic Press.

Bownes, M. and S. Pathirana. 2003. Comparative aspects of oogenesis. In: **Biology and Pathology of the Oocyte. Role in Fertility and Reproductive Medicine.** (A. Trounson and R. Gosden, eds.). Cambridge University Press.

Bronson, F. 1989. **Mammalian Reproductive Biology.** University of Chicago Press.

Byskov, A. and M. Nielson. 2003. Ontogeny of the mammalian ovary. In: **Biology and Pathology of the Oocyte. Role in Fertility and Reproductive Medicine.** (A. Trounson and R. Gosden, eds.). Cambridge University Press.

Byskov, A. 1982. Primoidial germ cells and regulation of meiosis. In: **Reproduction in Mammals. Germ Cells and Fertilization.** 2nd ed. (C. R. Austin and R. V. Short, eds.). Cambridge University Press.

Clark, I., Phillips, D., Mercer, J. and Robertson, D. 1991. Mechanisms involved in LH surge. In: **Neuroendocrine Regulation of Reproduction.** (S. Yen and W. Vale, eds.). Serono.

Dai, Y., J. Fulka and R. Moor. 2003. Checkpoint controls in mammalian oocytes. In: **Biology and Pathology of the Oocyte. Role in Fertility and Reproductive Medicine.** (A. Trounson and R. Gosden, eds.). Cambridge University Press.

Downs, S. 2002. Biochemistry of oocyte maturation. In: **The Future of the Oocyte. Basic and Clinical Application.** (J. Eppig, C. Hegele-Hartung and M. Lessl, eds.) Springer-Verlag.

Downs, S. 1995. Ovulation 2. Control of the resumption of meiotic maturation in mammalian oocytes. In: **Gametes: The Oocyte.** (J. Grudzinskas and J. Yovich, eds.). Cambridge University Press.

Eddy, E. and D. O'Brien. 1998. Gene expression during mammalian meiosis. In: **Meiosis and Gametogenesis.** (M. Handel, ed.). Academic Press.

Eppig, J., M.Viveiros, C. Marin-Bivens and R. de la Fuente. 2004. Regulation of mammalian oocyte maturation. In: **The Ovary.** 2nd edition. (P. Leung and E. Adashi, eds.) Elsevier Academic Press.

Findlay, J. 2003. Endocrine regulation of meiosis-local and systemic hormonal influences. In: **Biology and Pathology of the Oocyte. Role in Fertility and Reproductive Medicine.** (A. Trounson and R. Gosden, eds.). Cambridge University Press.

Firyal, S. and F. Khan-Dawood. 2003. The ovarian cycle. In: **Introduction to Mammalian Reproductive Physiology.** (D. Tulsiani, ed.). Kluwer Academic Publishers.

Fortune, J. 2002. Activation of primordial follicles. In: **The Future of the Oocyte. Basic and Clinical Application.** (J. Eppig, C. Hegele-Hartung and M. Lessl, eds.) Springer-Verlag.

Gordon, J. and L. Speroff. 2002. **Handbook of Clinical Gynecological Endocrinology and Infertility.** Lippincott, Williams and Wilkins.

171

SOURCES

Gosden, R. 1995. Oocyte development throughout life. In: **Gametes. The Oocyte**. (J. Grudzinskas and J. Yovich, eds.). Cambridge University Press.

Gougeon, A. 2003. The early stages of follicular growth. In: **Biology and Pathology of the Oocyte. Role in Fertility and Reproductive Medicine.** (A. Trounson and R. Gosden, eds.). Cambridge University Press.

Gould, J. and G. Gould. 1997. **Sexual Selection**. Scientific American.

Guraya, S. 2000. **Comparative Cellular and Molecular Biology of Ovary in Mammals**. Science Publishers, Inc.

Guyton, A and J. Hall. 2000. **Textbook of Medical Physiology.** 10th ed. W. B. Saunders.

Hafez, E. 1974. **Reproduction in Farm Animals.** 3 ed. Lea and Febiger.

Handel, M. and J. Eppig. 1998. Sexual dimorphism in the regulation of mammalian meiosis. In: **Meiosis and Gametogenesis.** (M. Handel, ed.). Academic Press.

Hunt, P. and R. LeMaire-Adkins. 1998. Genetic control of mammalian female meiosis. In: **Meiosis and Gametogenesis.** (M. Handel, ed.). Academic Press.

Hunter, R. 2003. **Physiology of the Graffian Follicle and Ovulation**. Cambridge University Press.

Hunter, R. 1988. **The Fallopian Tube**. Springer-Verlag.

Hyman, L. 1942. **Comparative Vertebrate Anatomy.** University of Chicago Press.

Izard, M. 1983. Pheromones and reproduction in domestic animals. In: **Pheromones and Reproduction in Mammals.** (J. Vandenburgh, ed.). Academic Press.

Jasienska, G. 2001. Why energy expediture causes reproductive suppression in women. In: **Fertile Grounds**. (P. Ellison, ed.). Aldine de Gruyter.

Jones, R. 1991. **Human Reproductive Biology.** Academic Press.

Karsch, F. 1984. The hypothalamus and the anterior pituitary. In: **Reproduction in Mammals. Hormonal Control of Reproduction.** 2 ed. (C. R. Austin and R. V. Short , eds.). Cambridge University Press.

Khan-Dawood, F. 2003. The ovarian cycle. In: **Introduction to Mammalian Reproductive Physiology**. (D. Tulsiani, ed.). Kluwer Academic Publishers.

Kosco, M. and J. Hill. 2003. **Principles of Mammalian Physiology**. Kosco Press.

Ledger, W. and D. Baird. 1995. Ovulation 3. Endocrinology of ovulation. In: **The Gametes: The Oocyte**. (J. Grudzinskas and J. Yovich, eds.). Cambridge University Press.

Mariana, J., Ronniaux, D., Draincourt, M. and Mauleon, P. 1991. Folliculogenesis. In: **Reproduction in Domestic Animals.** 4th ed. (P. T. Cupps, ed). Academic Press.

McClintock, M. 1983. Pheromonal regulation of the ovarian cycle: Enhancement, suppression and synchrony. In: **Pheromones and Reproduction in Mammals.** (J. Vandenburgh, ed). Academic Press.

Miller, G. 2000. **The Mating Mind**. Doubleday.

Mossman, H. and K. Duke. 1973. **Comparative Morphology of the Mammalian Ovary**. University of Wisconson Press.

Murphy, B. 2004. Luteinization. In: **The Ovary**. 2nd edition. (P. Leung and E. Adashi, eds.) Elsevier Academic Press.

Peters, H. and K, McNatty. 1980. **The Ovary. A Correlation of Structure and Function in Mammals**. University of California Press.

Pike, I. 2001. The evolution of ecological context of human pregnancy. In: **Reproductive Ecology and Human Evolution.** (P. Ellison, ed.). Aldine de Gruter.

Prasad, S., G. Kaul and B. Dunbar. 2003. **Structure and function of the mammalian zona pellucida.** In: **Introduction to Mammalian Reproductive Physiology**. (D. Tulsiani, ed.). Kluwer Academic Publishers.

Richards, J. 2001. Perspective:The ovarian follicle-A perspective in 2001. **Endocrinology 142:2184**.

SOURCES

Richards, J and Hendin, L. 1988. Molecular aspects of hormone action in ovarian follicular development, ovulation, and luteinization. **Annu. Rev. Physiol. 44:495.**

Richards, J. 1980. Maturation of ovarian follicles. Actions and interactions of pituitary and ovarian hormones on follicular cell differentiation. **Physiol. Rev. 60:51.**

Salustri, A., C. Fulop, A. Camaioni and V. Hascall. 2004. Oocyte-granulosa cell interactions. In: **The Ovary**. 2nd edition. (P. Leung and E. Adashi, eds.) Elsevier Academic Press.

Segal, S. 1989. The physiology of human reproduction. **Sci. Amer. 231:53.**

Short, R. 1994. Why sex? In: **The Differences Between the Sexes**. (R. Short and E. Balaban, eds.). Cambridge University Press

Shoubridge, E. 2003. The legacy of mitochondrial DNA. In: **Biology and Pathology of the Oocyte. Role in Fertility and Reproductive Medicine.** (A. Trounson and R. Gosden, eds.). Cambridge University Press.

Sole' R. and B. Goodwin. 2000. **Signs of Life**. Basic Books

Symons, D. 1979. **The Evolution of Human Sexuality**. Oxford University Press.

Tonetta, S. and DiZerega, S. 1989. Intragonadal regulation of follicular maturation. **Endocr. Rev. 10: 205.**

Wheater, P., Burkitt, H. and Daniel, V. 1979. **Functional Histology.** Churchhill and Livingston.

CHAPTER 5

Baird, D. 1984. The ovary. In: **Reproduction in Mammals. Hormonal Control of Reproduction.** 2nd ed. (C. R. Austin and R. V. Short, eds.). Cambridge University Press.

Baker, T. 1982. Oogenesis and ovulation. In: **Reproduction in Mammals. Germ Cells and Fertilization.** 2nd ed. (C. R. Austin and R. V. Short, eds.). Cambridge University Press.

Bedford, J. 1982. Fertilization. In: **Reproduction in Mammals. Germ Cells and Fertilization.** 2nd ed. (C. R. Austin and R. V. Short, eds.). Cambridge University Press.

Block, I. 1981. Sperm meets egg. **Science Digest 89:96.**

Cardullo, R. and C. Thaler. 2002. Function of the egg extracellular matrix. In: **Fertilization.** (M. Hardy, ed.). Academic Press.

Darszon, A., F. Espinoza, B. Galindo, D. Sanchez and C. Betran'. 2002. Regulation of sperm ion currents. In: **Fertilization**. (M. Hardy, ed.). Academic Press.

Eppel, D. 1977. The program of fertilization. **Sci. Amer. 237:128.**

Garbers, D. 1989. The regulation of spermatozoan function by the egg. In: **The Molecular Biology of Fertilization.** (H. Schatten and G. Schatten, eds.). Academic Press.

Gerton, G. 2002. Function of the sperm acrosome. In: **Fertilization**. (M. Hardy, ed.). Academic Press.

Gordon, J, and L. Speroff. 2002. **Handbook of Clinical Gynecological Endocrinology and Infertility**. Lippincott, William and Wilkins.

Harper, M. 1982. Sperm and egg transport. In: **Reproduction in Mammals. Germ Cells and Fertilization.** 2nd ed. (C. R. Austin and R. V. Short, eds.). Cambridge University Press.

Jaiswell, S. and M. Eisenbach. 2002. Capacitation. In: **Fertilization**. (M. Hardy, ed.). Academic Press.

Killan, G. 2003. Estrogen-associated glycoproteins in oviduct secretions: Structure and function for a role in fertilization. In: **Introduction to Mammalian Reproduction**. (D. Tulsiani, ed.). Kluwer Academic Publishers.

Kopf, G. 2002. Signal transduction mechanism regulating sperm acrosomal exocytosis. In: **Fertilization**. (M. Hardy, ed.). Academic Press.

Leibfreid, M., Florman, M. and First, N. 1989. The molecular biology of oocyte maturation. In: **The Molecular Biology of Fertilization.** (H. Schatten and G. Schatten, eds.). Academic Press.

SOURCES

Longo, F. 1997. **Fertilization.** 2nd edition. Chapman and Hall.

Longo, F. 1989. Egg cortical architecture. In: **The Cell Biology of Fertilization.** (H. Schatten and G. Schatten, eds.). Academic Press.

Ming, B., M. Wassler and M. Hardy. 2002. Sperm adhesion to the extracellular matrix of the egg. In: **Fertilization**. (M. Hardy, ed.). Academic Press.

Poccia, D. 1989. Reactivation and remodeling of the sperm nucleus after fertilization. In: **The Molecular Biology of Fertilization.** (H. Schatten and G. Schatten, eds.). Academic Press.

Primakoff, P. and D. Myles. 2002. Gamete fusion in mammals. In: **Fertilization**. (M. Hardy, ed.). Academic Press.

Quill, T. and D. Garber. 2002. Sperm motility activation and chemoattraction. In: **Fertility**. (M. Hardy, ed.). Academic Press.

Rories, C. and Spelsberg, T. 1989. Ovarian steroid action on gene expression. Mechanisms and models. **Annu. Rev. Physiol. 51:653**.

Ruiz-Braco, N. and Lennarz, N. 1989. Receptors and membrane interactions during fertilzation. In: **The Molecular Biology of Fertilization.** (H. Schatten and G. Schatten, eds.). Academic Press.

Short, R. 1984. Species differences in reproductive mechanisms. In: **Reproduction in Mammals. Reproductive Fitness**. 2nd ed. (C. R. Austin and R. V. Short, eds). Cambridge University Press.

Suarez, S. 2003. Transport of spermatozoa in the female genital tract. In: **Introduction to Mammalian Reproduction**. (D. Tulsiani, ed.). Kluwer Academic Publishers.

Suarez, S. 2002. Gamete transport. In: **Fertilization.** (M. Hardy, ed.). Academic Press.

Swann, A. and K. Jones. 2002. Membrane events and egg activation. In: **Fertilization**. (M. Hardy, ed.). Academic Press.

Thibault, C. 1973. Sperm storage and transport in vertebrates. **J. Reprod. Fert. 18:39**

Tienhoven, A. 1983. **Reproductive Physiology of Vertebrates.** 2 ed. Cornell University Press.

Tulsiani, D. and A. Abou-Haila. 2003. Sperm-egg interaction and exocytosis of acrosomal contents. In: **Introduction to Mammalian Reproduction**. (D. Tulsiani, ed.). Kluwer Academic Publishers.

Wasserman, P. 1988. Eggs, sperm and sugar. A recipe for fertilization. **News Physiol. Sci. 3:120.**

Wasserman, P. 1988. Fertilization in mammals. **Sci. Amer. 259:78**.

Wasserman, P. 1987. The biology and chemistry of fertilization. **Science 235:553**.

Westbrook, V., A. Diekman, J. Herr and P. Visconti. 2003. Capacitation: Signaling pathway involved in sperm acquisition of fertilizing capacity. In: **Introduction to Mammalian Reproduction**. (D. Tulsiani, ed.). Kluwer Academic Publishers.

CHAPTER 6

Anderson, G. 1991. Fertilization, early development and embryo transfer. In: **Reproduction in Domestic Animals.** 4th ed. (P. T. Cupps , ed.). Academic Press.

Arey, L. 1947. **Developmental Anatomy. A Textbook and Laboratory Manual of Embryology.** Saunders.

Balinsky, B. 1960. **An Introduction to Embryology**. Saunders.

Barry, J. 2002. Molecular Embryology. **How Molecules Give Birth to Animals**. Taylor and Francis.

Berrill, N. and G. Carp. 1976. **Development**. McGraw Hill.

Bier, E. 2000. **The Coiled Spring. How Life Begins**. Cold Spring Harbor Laboratory Press.

Bischof, P. P., A. Meisser and A. Campana. 2001. Biochemistry and molecular biology of trophoblast invasion. In: **Human Fertility and Reproduction. The Oocyte, the Embryo, and the Uterus. Annals of the New York Academy of Science, Volume 943.** (C. Bulletti, D. deZiegler, S. Guller and M. Levitz, eds.).

SOURCES

Bonner, J. 1988. **The Evolution of Complexity by Means of Natural Selection.** Princeton University Press.

Boving, B. 1971. Biomechanics of implantation. In: **The Biology of the Blastocyst**. (R. Blandeau, ed.). University of Chicago Press.

Coen, E. 1999. **The Art of Genes. How Organisms Make Themselves**. Oxford University Press.

Cohen, J. 1963. **Living Embryos**. Pergamon Press Ltd.

Cook, J. 1988. The early embryo and the formation of body pattern. **Am. Sci. 76:35**.

Cross, J. and S. Rossant. 2001. Development of the embryo. In: **Fetal Growth and Development**. (R. Harding and A. Bocking, eds.). Cambridge University Press.

Davies, J. and H. Hesseldahl. 1971. Comparative embryology of the mammalian blastocyst. In: **Biology of the Blastocyst.** (R. Blandau, ed.) University of Chicago Press.

Eakin, R. 1978. **Vertebrate Embryology**. 3rd edition. University of California Press.

Ebert, J. and I. Sussex. 1965. **Interactive Systems in Development.** 2nd edition. Holt, Rinehart and Winston, Inc.

Ellison, P. 2001. **On Fertile Ground. A Natural History of Human Reproduction.** Harvard University Press.

Enders, A. 1971. The fine structure of the blastocyst. In: **The Biology of the Blastocyst**. (R. Blandeau, ed.). University of Chicago Press.

Flood, P. 1991. The development of the conceptus and its relation to the uterus. In: **Reproduction in Domestic Animals.** 4th ed. (P. T. Cupps, ed.). Academic Press.

Gerhart, J. and M. Kirschner. 1997. **Cells, Embryos and Evolution**. Blackwell Science.

Gilbert, S. 1991. **Developmental Biology**. 3rd. edition. Sinauer.

Gilbert, S. 1989. **Pictorial Human Embryology**. University of Toronto Press.

Goodwin, B. 1976. **Analytical Physiology of Cells and Developing Organisms**. Academic Press.

Gould, S. 1977. **Ontogeny and Phylogeny**. The Belknap Press of Harvard University Press.

Grossinger, R. 1986. **Embryogenesis**. North Atlantic Books.

Hill, J. 2001. Maternal-embryonic cross-talk. In: **Human Fertility and Reproduction. The Oocyte, the Embryo, and the Uterus. Annals of the New York Academy of Science, Volume 943.** (C. Bulletti, D. deZiegler, S. Guller and M. Levitz, eds.).

Huettner, A. 1949. **Fundamentals of Comparative Embryology of the Vertebrates**. The MacMillian Company.

Langman, J. 1981. **Medical Embryology**. 4th edition. Williams and Wilkins.

Mathews, W. 1986. **Atlas of Descriptive Embryology**. MacMillian

McNamara, K. 1997. **Shapes of Time. The Evolution of Growth and Development**. The John Hopkins Press.

Mielczarek E. and S. McGrayne. 2000. **Iron, Nature's Universal Element. Why People Need Iron and Animals Make Magnets.** Rutgers University Press.

Moore, K. and T. Persaud. 2003. **Before We are Born. Essentials of Embryology and Birth Defects**. Saunders.

Mossman, H. 1971. Orientation and site of attachment of the blastocyst. In: **The Biology of the Blastocyst**. (R. Blandau, ed.). University of Chicago Press.

Nalbandov, A. 1971. Endocrine control of implantation. In: **The Biology of the Blastocyst**. (R. Blandau, ed.). University of Chicago Press.

Olsen, E. and R. Miller. 1958. **Morphological Integration**. University of Chicago Press.

Page, P. 1986. Metabolic aspects of the physiology of the preimplantation embryo. In: **Experimental Approaches to Mammalian Embryonic Development.** (J. Rossant and R. Pederson, eds.). Cambridge University Press.

Patten, B. 1929. **The Early Embryology of the Chick.** 3rd edition. P. Blackiston's Son and Co., Inc.

Raff, R. 1996. **The Shape of Life. Genes, Development, and the Evolution of Animal Form.** The University of Chicago Press.

SOURCES

Richa, J. and D. Solter. 1986. Role of cell surface molecules in mammalian development. In: **Experimental Approaches to Mammalian Embryonic Development.** (J. Rossant and R. Pederson, eds.). Cambridge University Press.

Rurak, D. 2001. Development and function of the placenta. In: **Fetal Growth and Development**. (R. Harding and A. Bocking, eds.). Cambridge University Press.

Sussman, M. 1973. **Developmental Biology. Its Cellular and Molecular Foundations**. Prentice-Hall, Inc.

Sussman, M. 1964. **Growth and Developent**. Prentice-Hall, Inc.

Thomson, K. 1988. **Morphogenesis and Evolution**. Oxford University Press.

Waddington, C. 1962. **How Animals Develop**. Harper Tourchbooks.

Waterman, A. 1948. **A laboratory Manual of Comparative Vertebrate Embryology**. Henry Holt and Co.

Weele, Cor van der. 1999. **Images of Development. Environmental Causes of Ontogeny**. State University of New York Press.

CHAPTER 7

Aynsley-Green, A. 1985. Metabolic and endocrine interrelations in the human fetus and neonate. **Am. J. Clin. Nutr. 41:399**.

Barbiera, R. 1994. The maternal adenohypophysis. In**: Maternal-Fetal Endocrinology.** 2nd edition. (D. Tulchinsky and A. Little, eds.). Saunders.

Beaconsfield, P., Birdwell, G. and Beconsfield, R. 1987. The placenta. **Sci. Amer. 243:94**.

Bissonnette, J. 1984. Regulation of cerebral blood flow in the fetus. **J. Dev. Physiol. 6:275**.

Boyd, R. and Y. Kudo. 1994. Transport functions of the placenta and fetal membranes. In: **Textbook of Fetal Physiology.** (G. Thornburn and R. Harding eds.). Oxford University Press.

Brace, R. 2001. Fluid balance. In: **Fetal Growth and Development.** (R. Harding and A. Bocking, eds.). Cambridge University Press.

Brace, R. 1994. Fetal fluid balance. In: **Textbook of Fetal Physiology.** (G. Thornburn and R. Harding eds.). Oxford University Press.

Brooks, N. 2001. Sexual development and development of reproductive organs. In: **Fetal Growth and Development.** (R. Harding and A. Bocking, eds.). Cambridge University Press.

Bruce, I. and J. Rawson. 1994. Development of motor functions of the fetus. In: **Textbook of Fetal Physiology.** (G. Thornburn and R. Harding eds.). Oxford University Press.

Byskov, A. 1986. Differentiation of the mammalian embryonic gonad. **Physiol. Rev. 66:71**.

Catchpole, H. 1991. Hormonal mechanisms in pregnancy and parturition. In: **Reproduction in Domestic Animals.** 4th ed. (P. T. Cupps, ed.). Academic Press.

Caveney, S. 1985. The role of gap junctions in development. **Annu. Rev. Physiol. 47:319**.

Challis, J., Kim, C., Naftolin, F., Judd, H., Yen, S. and Benirschke, K. 1974. The concentrations of androgens, oestrogens, progesterone and luteinizing hormone in the serum of fetal calves through the course of gestation. **J. Endocr. 60:107**.

Conley, A. and J. Mason. 1994. Endocrine function of the placenta. In: **Textbook of Fetal Physiology.** (G. Thornburn and R. Harding eds.). Oxford University Press.

Dawes, G. 1992. Fetal breathing in normoxia and in hypoxia. In: **Oxygen: Basis of the Regulation of Vital Functions in the Fetus.** (W. Kunzel and M. Kirchbaum, eds.). Springer-Verlag.

Falcone, T. and A. Little. 1994. Placental synthesis of steroid hormones. In**: Maternal-Fetal Endocrinology.** 2nd edition. (D. Tulchinsky and A. Little, eds.). Saunders.

Falcone, T. and A. Little. 1994. Placental peptides In**: Maternal-Fetal Endocrinology.** 2nd edition. (D. Tulchinsky and A. Little, eds.). Saunders.

SOURCES

Falcone, T. and A. Little. 1994. Maintenance of pregnancy and the initiation of normal and premature labor. In: **Maternal-Fetal Endocrinology.** 2nd edition. (D. Tulchinsky and A. Little, eds.). Saunders.

Fisher, D. and D. Poll. 1994. Development of the fetal thyroid system. In: **Textbook of Fetal Physiology.** (G. Thornburn and R. Harding eds.). Oxford University Press.

Fisher, D. and D. Polk. 1994. The ontogenesis of thyroid function. In: **Maternal-Fetal Endocrinology.** 2nd edition. (D. Tulchinsky and A. Little, eds.). Saunders.

Fowden, A. 2001. Growth and metabolism. In: **Fetal Growth and Development.** (R. Harding and A. Bocking, eds.). Cambridge University Press.

Fowden, A. 1994. Fetal metabolism and energy balance. In: **Textbook of Fetal Physiology.** (G. Thornburn and R. Harding eds.). Oxford University Press.

Froh, D. and P. Ballard. 1994. Fetal lung maturation. In: **Textbook of Fetal Physiology.** (G. Thornburn and R. Harding eds.). Oxford University Press.

Griffing, G. and J. Melby. 1994. The maternal adrenal cortex. . In: **Maternal-Fetal Endocrinology.** 2nd edition. (D. Tulchinsky and A. Little, eds.). Saunders.

Gluckman, P. 1994. Neuroendocrinology of the fetus. In: **Textbook of Fetal Physiology.** (G. Thornburn and R. Harding eds.). Oxford University Press.

Grumbach, M. and P. Gluckman. 1994. The human hypothalamus and pituitary: The maturation of the neuroendocrine mechanisms controlling the secretion of fetal pituitary growth hormone, gonadotropins, adrenocorticotropin-related peptides and thyrotropin. . In: **Maternal-Fetal Endocrinology.** 2nd edition. (D. Tulchinsky and A. Little, eds.). Saunders.

Han, V. and D. Hill. 1994. Growth factors in fetal growth. In: **Textbook of Fetal Physiology.** (G. Thornburn and R. Harding eds.). Oxford University Press.

Handson, M. and T. Kiserud. 2001. Cardiovascular system in fetal growth. In: **Fetal Growth and Development.** (R. Harding and A. Bocking, eds.). Cambridge University Press.

Harding, R. 1994. Development of the respiratory system. In: **Textbook of Fetal Physiology.** (G. Thornburn and R. Harding eds.). Oxford University Press.

Hay, W. and R. Wilkening. 1994. Metabolic activity of the placenta. In: **Textbook of Fetal Physiology.** (G. Thornburn and R. Harding eds.). Oxford University Press.

Head, R. and Flint, A. 1984. Pregnancy. In: **Reproduction in Mammals. Hormonal Control of Reproduction.** 2 ed. (C. R. Austin and R. V. Short, eds.). Cambridge University Press.

Heffner, L. 2001. **Human Reproduction at a Glance.** Blackwell Science.

Hooper, S. and R. Harding. 2001. Respiratory system in fetal growth. In: **Fetal Growth and Development.** (R. Harding and A. Bocking, eds.). Cambridge University Press.

Huch, A. and R. Huch. 1992. Maternal respiration: Its effect on the fetus. In: **Oxygen: Basis of Regulation of Vital Functions of the Fetus.** (W. Kunzel amd M. Kirchbaum, eds.). Springer-Verlag.

Itani, B. and R. Tsang. 1994. Calcium and mineral metabolism in the fetus. In: **Textbook of Fetal Physiology.** (G. Thornburn and R. Harding eds.). Oxford University Press

Jenkin, G. and P. Nathanielsz. 1994. Myometrial activity during pregnancy and parturition. In: **Textbook of Fetal Physiology.** (G. Thornburn and R. Harding eds.). Oxford University Press.

Johnson, M. and B. Everitt. 2000. **Essential Reproduction.** Blackwell Science.

Jones, C. and Rolph, T. 1986. Metabolism during fetal life. A functional assessment of metabolic development. **Physiol. Rev. 65:357.**

Josso, N. and Picard, J. 1986. Anti-mullerian hormone. **Physiol. Rev. 46:645.**

Kaplan, M. 1994. The maternal thyroid and parathyroid glands. In: **Maternal-Fetal Endocrinology.** 2nd edition. (D. Tulchinsky and A. Little, eds.). Saunders.

Kimpton, W., E. Washington and R. Cahill. 1994. Development of the immune system of the fetus. In: **Textbook of Fetal Physiology.** (G. Thornburn and R. Harding eds.). Oxford University Press.

King, D. and May, J. 1984. Thyroidal activity on body growth. **J. Exptl. Zool. 232:453.**

Kosco, M. and J. Hill. 2003. **Principles of Mammalian Physiology.** Kosco Press.

SOURCES

Lagercratz, H. and Slotkin, K. 1986. The "stress" of being born. **Sci. Amer. 254:100.**

Leake, R. 1994. The fetal-maternal neurohypophyseal system. In**: Maternal-Fetal Endocrinology.** 2nd edition. (D. Tulchinsky and A. Little, eds.). Saunders.

Liggins, G. 1994. The maternal endocrine system during pregnancy and parturition. In: **Textbook of Fetal Physiology.** (G. Thornburn and R. Harding eds.). Oxford University Press.

Lotgering, F. 1985. Maternal and fetal response to exercise during pregnancy. **Physiol. Rev. 65:1.**

Lumbers, E. 1987. Renal function during intrauterine life. **News Physiol. Sci. 2:220.**

Lye, S. and J. Challis. 2001. Parturition. In: **Fetal Growth and Development.** (R. Harding and A. Bocking, eds.). Cambridge University Press.

Marchlewski, A. 1983. Pregnancy blocking hormones. In: **Pheromones and Reproduction in Mammals,** (J. Vandenburgh, ed.). Academic Press.

Marx, J. 1973. Drugs and pregnancy. Do they effect the unborn child. **Science 180:174.**

Mielczarek E. and S. McGrayne. 2000. **Iron, Nature's Universal Element. Why People Need Iron and Animals Make Magnets.** Rutgers University Press.

Moore, D. and G. Jeffery. 1994. Development of the auditory and visual systems of the fetus. In: **Textbook of Fetal Physiology.** (G. Thornburn and R. Harding eds.). Oxford University Press.

Munro, H. 1983. The placenta in nutrition. **Annu. Rev. Nutr: 3:97.**

Nissely, S. and Rechler, M. 1984. Somatomedin/insulin-like growth factor tissue receptors. **Clin. Endocrinol. Metab. 13:43.**

Nuwayhid, B. and E. Quillen. 1994. Hypertension in pregnancy. In**: Maternal-Fetal Endocrinology.** 2nd edition. (D. Tulchinsky and A. Little, eds.). Saunders.

Olley, P. and Coceani, F. 1981. Prostiglandins and the ductus arteriosus. **Annu. Rev. Med. 32:375.**

Parkington, H. and H. Coleman. 1994. Regulation of contraction of uterine smooth muscle. In: **Textbook of Fetal Physiology.** (G. Thornburn and R. Harding eds.). Oxford University Press.

Perelman, R. 1985. Developmental aspects of lung lipids. **Annu. Rev. Physiol. 47:803.**

Pearson, B., P. Murphy and C. Branchard. 1994. The fetal adrenal. In**: Maternal-Fetal Endocrinology.** 2nd edition. (D. Tulchinsky and A. Little, eds.). Saunders.

Pelliniemi. L. and M. Dym. 1994. The fetal gonad and sexual differentiation. In**: Maternal-Fetal Endocrinology.** 2nd edition. (D. Tulchinsky and A. Little, eds.). Saunders.

Proske, U. 1994. Development of the skeletal muscle ant it innervations. In: **Textbook of Fetal Physiology.** (G. Thornburn and R. Harding eds.). Oxford University Press.

Rees, S. and D. Walker. 2001. Nervous and neuromuscular system. In: **Fetal Growth and Development.** (R. Harding and A. Bocking, eds.). Cambridge University Press.

Rees, S. 1994. The structural development of the nervous system. In: **Textbook of Fetal Physiology.** (G. Thornburn and R. Harding eds.). Oxford University Press.

Rigatto, H. 1984. Control of ventilation in the newborn. **Annu. Rev. Physiol. 46:661.**

Robillard., F. Smith, J. Seger, E. Guiller and P. Jose. 1994. Renal function in the fetus. In: **Textbook of Fetal Physiology.** (G. Thornburn and R. Harding eds.). Oxford University Press.

Rudolph, A. 1979. Fetal and neonatal pulmonary circulation. **Annu. Rev. Physiol. 41:383.**

Rurak, D. 2001. Development and function of the placenta. In: **Fetal Growth and Development.** (R. Harding and A. Bocking, eds.). Cambridge University Press.

Rurak, D. 1994. Fetal oxygenation, carbon dioxide homeostasis and acid-base balance. In: **Textbook of Fetal Physiology.** (G. Thornburn and R. Harding eds.). Oxford University Press.

Simpson, E. and MacDonald, P. 1981. Endocrine physiology of the placenta. **Annu. Rev. Physiol. 43:163.**

Smith, B. and M. Post. 1994. The influence of hormones on fetal lung development. In**: Maternal-Fetal Endocrinology.** 2nd edition. (D. Tulchinsky and A. Little, eds.). Saunders.

Sperling, M. 1994. Carbohydrate metabolism: Insulin and glucagon. In**: Maternal-Fetal Endocrinology.** 2nd edition. (D. Tulchinsky and A. Little, eds.). Saunders.

Thornburg, G. and M. Morten. 1994. Development of the cardiovascular system. In: **Textbook of Fetal Physiology.** (G. Thornburn and R. Harding eds.). Oxford University Press.

Thorburn, G. and Challis, J. 1979. Endocrine control of parturition. **Physiol. Rev. 59:863.**

Trahair, J. 2001. Digestive system in fetal growth. In: **Fetal Growth and Development.** (R. Harding and A. Bocking, eds.). Cambridge University Press.

Trahair, J. and R. Harding. 1994. Development of the gastrointestinal tract. In: **Textbook of Fetal Physiology.** (G. Thornburn and R. Harding eds.). Oxford University Press.

Walker, D. 1994. Development of the autonomic nervous system including adreno-chromaffin tissue. In: **Textbook of Fetal Physiology.** (G. Thornburn and R. Harding eds.). Oxford University Press.

Walker, D. 1984. Peripheral and central chemoreceptors in the fetus and newborn. **Annu. Rev. Physiol. 46:687.**

Weiss. G. 1994. The maternal ovaries. . In**: Maternal-Fetal Endocrinology.** 2nd edition. (D. Tulchinsky and A. Little, eds.). Saunders.

Weiss, G. 1984. Relaxin. **Annu. Rev. Physiology:46:43.**

Winter, J. 1994. The ontogeny of the pituitary-gonadal axis and sexual differentiation in the human. In: **Textbook of Fetal Physiology.** (G. Thornburn and R. Harding eds.). Oxford University Press.

Wood, C. 2001.Endocrine functions. In: **Fetal Growth and Development.** (R. Harding and A. Bocking, eds.). Cambridge University Press.

Wood, C. 1994. The function of the fetal-adrenal axis. In: **Textbook of Fetal Physiology.** (G. Thornburn and R. Harding eds.). Oxford University Press.

Yoshinga, K. 1994. Endocrinology of implantation. In**: Maternal-Fetal Endocrinology.** 2nd edition. (D. Tulchinsky and A. Little, eds.). Saunders.

CHAPTER 8

Catchpole, H. 1991. Hormonal mechanisms in pregnancy and parturition. In: **Reproduction in Domestic Animals.** 4th ed. (P. T. Cupps, ed.). Academic Press.

Dallman, J. 1981. Anemia of prematurity. **Annu. Rev. Med. 32:143.**

Ellison. P. 2001. **On Fertile Ground**. Harvard University Press.

Falcone, T. and A. Little. 1994. Maintenance of pregnancy and the initiation of normal and premature labor. In**: Maternal-Fetal Endocrinology.** 2nd edition. (D. Tulchinsky and A. Little, eds.). Saunders.

Guyton, A. and J. Hall. 2000. **Textbook of Medical Physiology.** 10th edition. Saunders.

Heffner, L. 2001. **Human Reproduction at a Glance.** Blackwell Science.

Jainudeen, M. and E. Hafez. 2000. Gestation, prenatal physiology and parturition. In: **Reproduction in Farm Animals**. 7th edition. Lippincott, Williams and Wilkins.

Johnson, M. and B. Everitt. 2000. **Essential Reproduction.** Blackwell Science.

Lagercratz, H. and Slotkin, K. 1986. The "stress" of being born. **Sci. Amer. 254:100.**

Lye, S. and J. Challis. 2001. Parturition. In: **Fetal Growth and Development.** (R. Harding and A. Bocking, eds.). Cambridge University Press.

Lye, S. 1994. The initiation and inhibition of labor: toward a molecular understanding. **Semin Reprod. Endocrinol. 12:284.**

Smith, R. 1999, The timing of birth. **Sci. Am. 280:65**

Thorburn, G. and Challis, J. 1979. Endocrine control of parturition. **Physiol. Rev. 59:863.**

Trevathan, W. 1987. **Human Birth**. Aldine de Gruyter

Walker, D. 1984. Peripheral and central chemoreceptors in the fetus and newborn. **Annu. Rev. Physiol. 46:687.**

SOURCES

CHAPTER 9

Aynsley-Green, A. 1985. Metabolic and endocrine interrelations in the human fetus and neonate. **Am. J. Clin. Nutr. 41:399.**

Burri, P. 1994. Structural development of the lung of the fetus and neonate. **In: Fetus and Neonate. Physiology and Clinical Applications. Breathing**. Volume 2. (M. Hanson, J. Spencer, C. Rodeck and D. Watters eds.) Cambridge University Press.

Durnian, J. 1984. Energy balance in childhood and adolescence. **Proc. Nutr. Soc. 43:271.**

Guyton, A. and J. Hall. 2000. **Textbook of Medical Physiology.** 10th edition. Saunders

Heffner, L. 2001. **Human Reproduction at a Glance.** Blackwell Science.

Heymann, M. 1981. Factors effecting changes in neonatal systemic circulation. **Annu Rev. Physiol. 43:371.**

Jameson, E. 1988. **Vertebrate Reproduction.** John Wiley and Sons.

Johnson, M. and B. Everitt. 2000. **Essential Reproduction.** Blackwell Science.

Mortola, J. 2001. **Repiratory Physiology of Newborn Mammals.** The John Hopkins University Press.

Mortola, J. 1994. Neonatal respiratory mechanics. In: **Fetus and Neonate. Physiology and Clinical Applications. Breathing**. Volume 2. (M. Hanson, J. Spencer, C. Rodeck and D. Watters eds.) Cambridge University Press.

Mortola, J. 1987. Dynamics of breathing in newborn mammals. **Physiol. Rev.67:187.**

Mimouni, I. and R. Tsang. 1994. Perinatal mineral metabolism In: **Maternal-Fetal Endocrinology.** 2nd edition. (D. Tulchinsky and A. Little, eds.). Saunders.

Oswald, P. and Peltman, P. 1974. The cry of the human infant. **Sci. Amer. 230:84.**

Read, D. and Henderson-Smart, D. 1984. Regulation of breathing in the newborn during different behavioral states. **Annu Rev. Physiol. 46:675.**

Rudolph, A. 1979. Fetal and neonatal pulmonary circulation. **Annu. Rev. Physiol. 41:383.**

Sperling, M. 1994. Newborn adaptation: Adrenal-cortical hormones and adrenocorticotropic hormone. metabolism In: **Maternal-Fetal Endocrinology.** 2nd edition. (D. Tulchinsky and A. Little, eds.). Saunders.

Thornburg, K. and M. Morten. 1993. Growth and development of the heart. **In: Fetus and Neonate. Physiological and Clinical Applications. The Circulation.** Volume 1. (M. Hanson, J. Spencer and C. Rodeck, eds.) Cambridge University Press.

Walker, A. 1993. Circulatory transitions at birth and the control of neonatal circulation. **In: Fetus and Neonate. Physiological and Clinical Applications. The Circulation.** Volume 1. (M. Hanson, J. Spencer and C. Rodeck, eds.) Cambridge University Press.

Walker, A. 1984. Peripheral and central chemoreceptors in the fetus and newborn. **Annu. Rev. Physiol. 46:687.**

Wyatt, J., A. Edwards and O. Reynolds. 1993. Assessment of cerebral hemodynamics and oxygenation in newborn human infant. **In: Fetus and Neonate. Physiological and Clinical Applications. The Circulation.** Volume 1. (M. Hanson, J. Spencer and C. Rodeck, eds.) Cambridge University Press.

CHAPTER 10

Baldwin, R. and Miller, P. 1991. Mammary gland development and lactation. In: **Reproduction in Domestic Animals.** 4th ed. (P. T. Cupps, ed.). Academic Press.

Carpenter, G. 1981. The importance of mothers milk. **Natural History Magazine. 9:6.**

Cowie, A. 1984. Lactation. In: **Reproduction in Mammals. Hormonal Control of Reproduction.** 2 ed. (C. R. Austin and R. V. Short, eds.). Cambridge University Press.

SOURCES

Cowie, A. and J. Tindel. 1971. **The Physiology of Lactation**. William and Wilkins Co.

Ganong, W. 1991. Role of the nervous system in reproduction. In: **Reproduction in Domestic Animals.** 4 ed. (P. T. Cupps, ed.) Academic Press.

Gaul, G. 1984. Significance of growth modulators in human milk. **Pediatrics 75:142**.

Guyton, A. and J. Hall. 2000. **Textbook of Medical Physiology**. 10th edition. Saunders.

Heffner, L. 2001. **Human Reproduction at a Glance.** Blackwell Science.

Jameson, E. 1988. **Vertebrate Reproduction.** John Wiley and Sons.

Johnson, M. and B. Everitt. 2000. **Essential Reproduction**. Blackwell Science.

Kosco, M. and J. Hill. 2003. **Principles of Mammalian Physiology**. Kosco Press.

Miller, G. 2000. **The Mating Mind**. Doubleday.

Nandi, S. 1959. Hormonal control of mammogenesis and lactogenesis in C3H/HE Crg/mouse. **Univ. Calif. Publs. Zool. 65:1**.

Peaker, M. and Wilde, C. 1987. Milk secretion. Autocrine control. **New Physiol. Sci. 1:131**.

Squires, E. 2003. **Applied Animal Endocrinology**. CABI Publishing.

Tucker, H. 2000. Hormones, mammary growth and lactation. **J. Diary Sci. 83:874.**

Vonderhaar, B. and Ziska, S. 1989. Hormonal regulation of milk protein gene expression. **Annu. Rev. Physiol. 51:641.**

INDEX